Fish Sticks, Sports Bras, and Aluminum Cans

FISH STICKS, SPORTS BRAS, AND ALUMINUM CANS

The Politics of Everyday Technologies

Paul R. Josephson

Johns Hopkins University Press

BALTIMORE

Johns Hopkins University Press
2715 North Charles Street
Baltimore, Maryland 21218-4363
www.press.jhu.edu

Library of Congress Cataloging-in-Publication Data

Josephson, Paul R.
    Fish sticks, sports bras, and aluminum cans : the politics of
everyday technologies / Paul R. Josephson.
        pages cm
    Includes bibliographical references and index.
    ISBN 978-1-4214-1783-7 (pbk. : alk. paper) — ISBN 1-4214-1783-9
(pbk. : alk. paper) — ISBN 978-1-4214-1784-4 (electronic) —
ISBN 1-4214-1784-7 (electronic)   1. Technological innovations—
Social aspects.   2. Popular culture.   I. Title.
    GN406.J57 2015
    303.48'3—dc23   2014049514

A catalog record for this book is available from the British Library.

*Special discounts are available for bulk purchases of this book. For more
information, please contact Special Sales at 410-516-6936 or specialsales@
press.jhu.edu.*

Johns Hopkins University Press uses environmentally friendly book
materials, including recycled text paper that is composed of at least
30 percent post-consumer waste, whenever possible.

From ghoulies and ghosties
And long-leggedy beasties
And things that go bump in the night,
Good Lord, deliver us!
Traditional Scottish Prayer

Contents

Acknowledgments

I would like to thank the students of "Luddite Rantings" at Colby College for listening to some of these thoughts in lecture form and suggesting topics for analysis, and especially to my research assistants Carrie Ngo, Kaitlin McCafferty, Charlie Spatz, Melissa Meyer, and Syd Hammond. Melanie Arndt, Marion Nestle, and Michael Gordon offered comments or suggestions at some point in the process. Bob Brugger at Johns Hopkins University Press helped turn my idea for this book into a book. Colby College and the Rachel Carson Center provided support for this book; my colleagues in the history department at Tomsk State University also contributed to this study. Charlotte Wilder admitted that she shares my fascination with reading cereal box contents. Indiana Jones is a fine scholar and great indexer. My former high school colleagues in the Mt. Lebanon Fortieth Anniversary Philosophy Club contributed to this book through weekly discussions about the meaning of technology, consumption, and politics. Willie Stargell and Roberto Clemente continue to provide inspiration, as do members of the OMTC: Erik, Jon, and Darren. I dedicate this book to Tanya Kasperski for having listened to my arguments about technology, helping me clarify many of them, and running my 100th marathon in Paris with me.

Fish Sticks, Sports Bras, and Aluminum Cans

Empty soft drink bottles in cases, Mapusa, Goa, India. Bottles and cans provide the world with more than two billion servings of surgery liquid daily. Photo by Katherine Goryunova.

Technostories

What do aluminum cans, sports bras, rocket ships, and French fries have in common? Over the past few years I have pondered the histories of these modern technologies in search of common themes and lessons. I am interested both in everyday objects (bananas and sports clothing) and in those considered the epitome of the modern scientific world (airliners and nuclear reactors). I am intrigued by the dictum of the nineteenth-century German philosopher Ludwig Feuerbach ("Der Mensch ist, was er isst"—or, loosely, "You are what you eat," and therefore fish sticks and high fructose corn syrup) and also by the claims of world leaders about the inherent goodness of achievements in science and technology and confidence in going forward, no matter the costs (George W. Bush on going into the cosmos, "We do not know where this journey will end" or Vladimir Putin on one of his pet projects, the Sukhoi jet, "I know, I've flown").

In each case, inventors and capitalists, colonial rulers and subjects, workers and managers, peasant farmers and investment bankers, consumers and regulators, elected public officials and corporate CEOs have contributed to the rise of modern technology. My understandings about technology resemble those of Thomas Hughes, David Nye, and others about technological systems.[1] Hughes offers a definition of technological systems as consisting of physical components (artifacts), natural resources, and organizations (manufacturing firms, utility companies, investment banks, scientific and engineering research organizations, and universities) that interact with other artifacts. System builders and their associates bring the systems on line through difficult processes of invention, development, innovation, and technology transfer; through fund-raising and public relations; and running through regulatory, political, and other obstacles. In other words, we cannot look at objects in isolation, but must consider the messy interaction of engineering, scientific, financial, governmental, consumer, and social institutions in giving impetus—or creating obstacles—to the dissemination of technology, and we must recognize all of the actors involved in their history.

I use technologies familiar to us to explore how they developed, how they were marketed and advanced, and how they are shaped by society and culture. Yet, as David Nye reminds us, many technologies "become more rigid and less responsive to social pressures" once ownership, control, and technological specifications are established, and once vested interests begin to insist on ways to shape and direct the system.[2] We often see this in the behavior of institutes, research centers, corporations, trade associations, and government regulators. This is the phenomenon known as technological momentum,[3] for example, when trade groups find it difficult to wean themselves from government largesse and oppose regulations that may require changes.

Each of the technologies considered in this book is a technological system, not an object-in-itself, although I may treat it as such from time to time. The aluminum can is bauxite mining operations that developed during the Cold War. Multinational corporations gained a foothold in Jamaica with the assistance of the colonial government to mine the ore with both expected and unanticipated impacts on the peasantry, employment, environment, and governance. The sports bra grew from the efforts of two female inventors who joined ideas of gender equality to advances in materials science and to the increasing numbers of female athletes whose participation owed something to Title IX antidiscrimination rulings, not only because of consumer demand. The fish stick arose in part because of a glut caused by technological overfishing, not consumer demand.

My approach here is not intended to be deeply theoretical, but rather, I hope, engaging and readable, although I raise crucial issues in each chapter and offer suggestions for further reading for each chapter at the end of the book. Based on primary source research in scores of newspapers; in scientific journals on food and freezing, materials science, physiology, engineering and so on; in archives; and extensive field research, I identify common themes in the history of nineteenth- and twentieth-century technology. The material ranges in geography from Jamaica to the North Atlantic fisheries; from exercise gyms to congressional offices; from natural disasters in South Asia to flooding along the

Mississippi River. It considers the lives and interests of fishermen, peasant farmers, marathon runners, migrant workers, lobbyists, scientists, and engineers. I organize each chapter around a series of technologies connected to one major artifact and also to a series of larger issues—participation, equity, health and safety, consumer demand, and so on—with the suggested readings to examine those larger issues in greater detail. Taken together, the short essays may serve as a textbook for a high school or college course, or individually, may illustrate important points about the place of technology in the modern world. In six chapters I situate a series of familiar technologies in broader economic, political, and cultural contexts: technology and empire; technology and consumerism; technology, gender, and sport; technology and state power; technology and the imperative for mass production; and technology and nature.

Each chapter concerns a specific technological system or iconic symbol of that technology and can be read on its own. But the reader will find common points of interest and concern across chapters. The chapter on fish sticks analyzes the rise of the supermarket, the freezer display case, and the "modern" postwar kitchen with such space-age appliances as the modern stove and later microwave oven in which the housewife could prepare "ready-to-eat" foods, and thus raises issues of gender and family life, while discussion of the sports bra also encourages the reader to engage a series of concerns about gender and technology.

Environmental concerns have a place in several chapters. In "fish sticks" we consider briefly overfishing of cod and other white fish in the North Atlantic Ocean. When analyzing the rise of potato and corn monocultures in the Pacific Northwest and the plains states, respectively, environmental issues pertaining to land use and chemical inputs come into focus. Of course, the chapter on natural disasters and technology keeps environmental issues front and center. The transformation of the Everglades into a sugar factory and the Columbia River into an electricity, fruit, and vegetable farm, the impact of bananas and bauxite mining on Jamaican ecosystems, and so on—these events and phenomena tell us that human-nature-technology interactions in the twentieth century reflect attitudes about power, progress, and the desirability—or dangers—of environmental change.

Food technologies—the French fry, the dehydrated potato, corn sweeteners, and fish sticks—play a central role in several of our discussions, not the least for direct and indirect federal subsidies to foods, the central place of foods in the massive, national school lunch program, and issues of public health, regulation, and labeling including the safety of school lunches. What kind of relationships do agribusinesses, their trade associations, and school lunches have, and how should we ensure that children receive healthy lunches given the domination of our table by processed foods?

Several of the chapters discuss consumer demand, technology, and desire, much of which is manipulated by wealthy corporations and their trade associations. Several of them address the politics and geopolitics of technology in colonialism, postcolonialism, and the Cold War (the chapter on Jamaica, bananas, and bauxite and the chapter on state power and technological symbolism).

In chapter 1, initially published in *Technology and Culture*, I explore fish sticks, the "ocean's hot dog." I examine how Americans (and Europeans) created the fish stick less because of consumer demand and more in response to overproduction based on new fishing and refrigeration technologies, including sonar that enabled locating schools of fish, advances in materials science that led to light, strong, and very large nets that allowed for bigger catches, a weak fish market, and the rise of the postwar kitchen. At the same time, fish processors, in order to market their portion-sized meals, took advantage of the rise of the modern kitchen and the modern housewife, who signaled her modernity in part through the preparation of "heat and serve" prefabricated meals. Government regulation, in determining that "fish" sticks must have at least 60 percent "fish matter" in them, also played a role—thankfully—in determining the fish essence of the fish stick. But it was certainly not the great clamoring masses who insisted that the gods of food preparation bring them the ocean's hot dog.

As an avid marathoner, a writer for a runners' magazine, and a noncompensated sports equipment tester for a running shoe and apparel company, I became fascinated with how changes in technology had an impact on sports performance, and in particular with gender and technology. Drawing on interviews with pioneers in the technology and a rich literature in physiology, kinesiology, and other journals, I fashioned a history of the sports bra. The sports bra (chapter 2) represents a surprising story of "reverse gender engineering," plus advances in materials science, a revolution in female sports participation, and consumer demand generated by federal antidiscrimination rulings. The inventors of the first sports bra engineered it from two jock straps. This contributed to cross-strap designs in the back and an understanding that all seams must be on the outside and enabled women to be active jocks themselves. The advances in materials science (Lycra, spandex, wicking material) created support with comfort and contributed to a lightweight design that was also an attractive and modest fashion item. And Title IX rulings required equal opportunity for women in education and sports, meaning that suddenly millions of woman athletes—and jazzercisers and others—demanded this new sports equipment.

Food is a central concern in chapter 3 on technology, colonialism, and post-colonialism, in this case the role of sugar, then bananas, and finally aluminum in international trade. Slavery was used to exploit Caribbean societies, and later various other land-grab schemes kept the emancipated peasant weak and vulnerable. Using the aluminum soda can as an icon of the Anglo-European sweet tooth, I focus on the Jamaican experience. At each step of the exploitation of Jamaican resources, the use of technologies of transport, agriculture, mining, and infrastructure lagged or advanced depending on such broader sociopolitical factors as slavery and its impact on technological innovation; the rise of banana trade, banana monocultures, and banana republics; and the continued colonial usurpation of resources with the Cold War determination to extract bauxite ore from the island.

I also have a sweet tooth. And I seem to be addicted, unfortunately, to a store-bought chocolate chip cookie with high fructose corn syrup as a major ingredient. The industrial paradigm and the headlong drive to standardization of food production is the focus of chapter 4, where we explore industrial potatoes and the industrial sweetener high fructose corn syrup (HFCS) ("Mass-Produced Nutrition"). Both potatoes and HFCS grew out of the seemingly self-augmenting and autonomous development of a series of technological systems; this is an argument known as "technological determinism," a major discussion of which appears in Jacques Ellul's *The Technological Society* (1964). The point is that hydroelectricity, irrigation, modern shipping, and monocultures come together in agribusinesses. The agribusinesses and their trade associations vigorously defend the right to sell mass-produced foods whose nutritional value and health risks are subject to continuing debate.

The increased popularity of the Weather Channel and the effort of various media to sell stories to generate viewership—rather than report the news and provide analysis of events—has led to intensive coverage of natural disasters and the fury of "mother nature," especially the closer those disasters are to home (e.g., Hurricanes Katrina and Sandy). As a nuclear historian, I wondered what factors connected such disasters—hurricanes, floods, tsunamis, earthquakes, Fukushima, Chernobyl, and so on. Chapter 5 ("Technology and (Natural) Disasters: You Cannot Fool (Mother) Nature") debunks the notion that natural disasters exist; they are human and technological and social and natural all at the same time. We begin "in nature," in Fukushima Prefecture in Japan, a coastal region known for its rich fishing industry, whose upland regions are agricultural. The story involves a large-scale, self-augmenting technology: the Tokyo Electric Power Company, which joined with General Electric in 1971 to build six boiling water reactors with a total capacity of 4,700 megawatts on the seashore. How could engineers, power companies, and government regulators permit the construction of nuclear power stations along the beaches of the Pacific Ocean in a region of agriculturalists and fishermen, and one of great seismic activity where a massive tsunami was a likelihood?

The Fukushima nuclear disaster led many people to ponder natural disasters and triggered a rebirth of "disaster history." Fukushima joined the lexicon of world technological disaster history along with Bhopal, *Exxon Valdez*, Chernobyl, and other accidents. Can we or should we separate these events from such natural disasters as floods, Hurricanes Katrina and Sandy, and forest fires, including those in Russia in 2010 that, along with oppressive heat and thick, dangerous smoke, may have killed tens of thousands of people? How do sociocultural and economic issues shape government response to disaster? What kinds of regulations might limit their impact? Admittedly, some disasters seem to be more "natural"—a hurricane or monsoon, earthquake or volcano, for example, and others more "human" or "anthropogenic," for example, an oil refinery fire, an oil tanker running aground, or a reactor explosion. Yet whether disasters arise from cumulative causes or understandable but unpredictable ones, both human and natural agency play a role in their occurrence. We should avoid viewing disasters in binary fashion, but rather as something complex that involve humans, their technological systems, and nature. Simply put, without humans to apprehend nature, there are no natural disasters. Indeed, natural disasters involve human presence—villages, homes, cemeteries, businesses, factories, schools, and hospitals, which are often situated in floodplains or on the coast, for example. These are often built in such locales because of such perverse incentives as federally subsidized insurance, and because of such federally funded science and engineering as flood abatement projects—as I explore in the history of disasters involving infrastructure on the floodplain; earthquakes, schools, apartment buildings; and engineering hubris (supertankers, and reactors).

I return to my frequent focus on the former Soviet Union in chapter 6, "Big Artifacts: Technological Symbolism and State Power." Russian President Vladimir Putin loves to be photographed shirtless, in a natural setting. His handlers intend these publicity photos to show him, like Joseph Stalin, to be more powerful than nature itself. Putin has also appeared in nature fully clothed in a white jumpsuit, piloting a hang glider in an attempt to lead endangered Siberian cranes into the wild. The "Flight of Hope," as it was called, left the cranes nonplussed; they dropped out and returned to captivity on a passenger plane. But, as his technological feats along the Northern (Arctic) Sea Route, in nuclear power, aerospace, and skyscrapers, along with his rhetoric and photo sessions reveal, Putin is determined to lead Russian citizens into the glorious future, one in which Russia is a superpower. These technologies of government legitimacy resemble in many ways those built in the 1930s, 1940s, and 1950s: Stalin's hydropower stations, "Wedding Cake" skyscrapers, and the military-industrial complex.

Russia is not alone in the effort of handlers, journalists, and propagandists to demonstrate the glories of state power and legitimacy on the foundation of technological achievements in aerospace, skyscrapers, highways, and other large-scale technological systems. Think of the importance of "Atoms for Peace" programs worldwide after President Dwight Eisenhower's call to domesticate the atom at the United Nations in 1953, or President John Kennedy's speech in 1961 announcing that the United States would win the Space Race. Think of technological fairs of the nineteenth century. Think of Gothic cathedrals. The symbolism of big technology is a subject for all centuries and all cultures.

Unloading cod from a modern trawler. Twentieth-century advances in fishing, processing, and refrigeration enabled dramatic increases in production in the commercial fishing industry. Courtesy of National Oceanographic and Atmospheric Administration, National Marine Fisheries Service. Photo by Allen M. Shimada.

1

The Ocean's Hot Dog

The Development of the Fish Stick

Der Mensch ist, was er isst.
[Loosely, "You are what you eat."]
—LUDWIG FEUERBACH

The fish stick—the bane of schoolchildren who generally consider it an over-cooked, breading-coated, cardboard-tasting, fish-less product of lunchrooms and of mothers to deceive them into consuming protein—is a postwar invention that resulted from the confluence of several forces of modernity. These forces included a boom in housing construction that contained kitchens with such new appliances as freezers; the seeming appeal of space-age, ready-to-eat foods; the rise of consumer culture; and an increasingly affluent society. Yet the fish stick appeared during the 1950s not because consumers cried out for it, and certainly not because schoolchildren demanded it, but because of the need to process and sell tons of fish that were harvested from the ocean, fil-leted, and frozen in huge, solid blocks. Consumers were not attracted by the form of these frozen fillets, however, and demand for fish products remained low. Manufacturers believed that the fish stick—a breaded, precooked food—would solve the problem. Still, several simultaneous technological advances had to take place before the product could appear in your grocer's freezer.[1] These advances occurred in catching, freezing, processing, and transportation technologies.

The postwar years witnessed a rapid increase in the size of merchant ma-rines in many countries, with these merchant fleets adopting new, almost rapa-cious catching methods and simultaneously installing massive refrigeration and processing facilities onboard huge trawlers. Sailors caught, beheaded, skinned, gutted, filleted, and then plate- or block-froze large quantities of cod, pollock, haddock, and other fish—tens of thousands of pounds—and kept them from spoiling in huge freezing units. Once on shore, the subsequent attempt to separate whole pieces of fish from frozen blocks resulted in mangled, unappetizing chunks. Frozen blocks of fish required a series of processes to transform them into a saleable, palatable product. The fish stick came from fish blocks being band-sawed into rectangles roughly three inches long and one inch wide (~7.5 × 2.5 cm), then breaded and fried. Onboard processors eventually learned to trim fish into fillets and other useable cuts before freezing. Processors considered these other cuts the "portion," which found a home in institutional kitchens (schools, hospitals, factories, and restaurants). Fish sticks had a largely retail success because demand for them in schools and elsewhere waned as more manufacturers entered production and quality declined.

How Gorton's, based in Gloucester, Massachusetts, entered the fish stick market and achieved a leading position is the story of this chapter. The information, which is based on corporate archives and industry publications, focuses on supply-side factors that contributed to the rise of the fish stick as an important icon of US food-product ingenuity. I focus on Gorton's for two reasons. First, the company was a pioneer in the portion and fish stick industry and has remained at the cutting edge of product innovations in institutional and home products, and, along with the Birds Eye and Mrs. Paul's companies, has dominated the fish stick industry in sales from the beginning.[2] And second, I believe the Gorton's experience with the fish stick is paradigmatic of the industry. Materials from its corporate archives reveal clearly how technology, marketing, and other forces led to the invention of the fish stick. I do not intend this chapter to be a paean to Gorton's; the company was, however, a leader in product development and maintained higher levels of quality control than many other manufacturers.

Consumer demand, consumer attitudes, changes in the postwar American household and family also contributed to the success of the fish stick. But its success had more to do with the revolutions in catching, processing, and preparing frozen foods, along with other factors. One of these was an apparently successful marketing campaign directed at busy housewives; another was the role of the federal government in developing, promoting, and regulating new food products and in providing markets for them through school lunch programs. University scientists—in the case of Gorton's, those at the Massachusetts Institute of Technology (MIT)—gave rise to the modern fish stick through research funded jointly by the US government and Gorton's. This research, which was tied to the expansion of supermarkets and the refinement of refrigeration, processing, shipping, and display of products, also fostered the creation of such products as the TV dinner. Gorton's, Birds Eye, and others showed the way with the fish stick, a product made of grade-A fish, light breading, and a few additives.

The Technology of Freezing, Packing, and Catching

The fish stick grew out of a half century of innovations in food preservation techniques. Salt and other additives used to prolong products' shelf life gave way to canning, refrigeration, and freezing so that unspoiled and wholesome fruit, vegetables, meat, and fish reached the consumer. Several individuals, the most well-known of whom was Clarence Birdseye, contributed to "quick freezing" processes and packaging innovations such as a moisture-proof cellophane wrapping. Quick freezing and other new processes rendered frozen products more palatable to consumers.[3] Early on, government and private researchers focused on freezing vegetables and fruits, not fish.[4] The first attempts to provide consumers with fresh or frozen fish using new refrigeration technologies failed; the fish had refrigeration burns, a tough texture, and they often smelled, and gills, stomach contents, and slimy skin frequently incubated bacterial infections. Refrigeration only retarded spoilage and had to be accomplished as quickly as possible after the fish were caught and cleaned—storing them on ice was not enough. Scientists therefore sought to combine refrigeration and freezing with various chemical dips. They experimented with mild antiseptics that were considered harmless to humans but killed fish bacteria, and poisons such as chlorine were even added to the ice itself. Soaking in brine with other chemicals was another possibility, although this led to the deterioration in appearance of the fish. Such antioxidants as ascorbic acid applied in glazes and additives were also effective in preventing oxidative rancidity (spoilage) of fish during storage. As in other areas of industrial food production, fishery specialists explored the use of antibiotics to deal with spoilage—though this was no panacea because failure to ice the fish promptly and properly after they were caught seriously interfered with antibiotics' effectiveness.[5]

Freezing technology is relatively simple: fans blow cold air over banks of finned coils, and this air freezes the products passing near the coils, often on conveyor belts in large assembly line–like facilities. Yet the freezer design of freezing had several negative features relative to fish processing. Air circulation in freezers, combined with a slight drop in vapor pressure when water from the fish condensed on the coils, accelerated fishes' dehydration. Researchers determined that a critical velocity of the dry, cold air existed, above which the fish were damaged: the higher the air temperature, the lower the air velocity needed to cause damage.[6] Researchers also established that freezing started in intercellular regions where salt concentration was the lowest, which resulted in intercellular fluids becoming more concentrated than fluids within cells. Water then left cells through osmosis, especially during extremely slow freezing; this "drip" (or dehydration) resulted in dry, tasteless meats, vegetables, and fish of low nutritional value that often had an "off taste." Quick freezing overcame this problem partly because only small ice crystals formed in the fish, thus allowing the cells to remain intact.[7]

Like other agricultural- and food-industry technological changes, refrigeration and freezing ended the seasonality of many products. This enabled the food industry to meet the demand for fish and other products during winter months when supplies were at their lowest.[8] Yet, for many reasons, the time was not right for such "heat and eat" products: during the Great Depression, high unemployment and low incomes discouraged development of new food products; limited trawling and refrigeration capacity prevented the economical production, distribution, and sale of frozen seafood products; packaging was unattractive; and last, many of the products simply did not catch on. For example, the fishing industry tried marketing "fishbricks," which were quick-frozen filleted fish packaged like blocks of ice cream. The main selling point was that "the housewife can cut the fish into any shape and be confident that the shape will be retained even after cooking"; no defrosting was necessary. But the First National and Kroger's grocery stores could hardly sell the product, and moreover, most stores lacked frozen-food display cases to accommodate the bricks.[9]

The market began to change when General Foods, using the Birds Eye trademark, fostered the development of freezer cabinets in grocery stores.[10] Birdseye had first marketed his quick-frozen foods in 1930 in Springfield, Massachusetts, and thereafter hundreds of food processors moved quickly into the frozen-food industry and introduced thousands of products. On the basis of Birds Eye's technology, Thompson Spa introduced main-course dinners in 1938; Mrs. Paul's and Timeliness joined in, as did Swift and Sara Lee with such ready-to-heat foods as pot pies. By 1945, there were 450 quick-freezing firms selling 600 million pounds of frozen foods through 40,000 retail outlets. War rationing of canned goods for the armed forces led even more customers to frozen foods. Supermarkets expanded their freezer display cases to meet such rising demand.[11]

The fish stick also prospered from changes in catch and transport technologies. Increases in the size and speed of trawlers and their rapid handling of products resolved the problems of freshness and economy. Small boats that stayed close to shore gave way to large, powerful vessels that might trawl for days at a time, often working in groups that sold ever larger catches to canning and other processing factories along the coasts. Nets made of durable, flexible materials (eventually plastics) made it possible to catch almost indiscriminately, and increasingly powerful winches hauled in nets bursting with sea life. Oceanographic data on water chemistry, currents, and fish populations and migration behavior enabled captains to find schools of fish more easily. Ultimately, such military innovations as sonar made it possible to locate potential catches with little delay.

A crucial innovation for the fish stick was the construction of vessels with huge capacities for refrigeration and freezing. After World War II, the nations of the North Atlantic launched floating factories and trawlers that froze catches at sea. New trawlers had storage capacities of 15,000 tons and more. Trawlers often worked in pairs, pulling nylon nets that were kilometers in length, hauling in the fish living on or near the bottom along with pelagic fish (those that live in the open seas) and freezing them according to size, species, and other criteria.

International competition contributed to the taking of massive catches and over-fishing by the new trawlers and so-called floating factories, the fleets of many North Atlantic countries seeking out cod, haddock, and other fish. Most governments subsidized their fleets; for example, the Canadian government subsidized modern refrigeration plants and trawlers to provide steady employment for coastal fishermen in the maritime provinces.[12] Its support extended to Quebec province in the construction of a modern freezer and filleting plant at Grindstone on the Magdalen Islands, which supplied Gorton's with much of its catch. The US government belatedly got into the act of subsidizing trawler manufacture in the late 1960s,[13] and by then fleets of trawlers had already depleted nearby fishing grounds such as George's Bank off the New England coast. To ensure a steady supply of fish, Gorton's was forced to enter world markets with its own and also contracted vessels; to this day, Gorton's contracts with the Polish Distant Water Fleet.[14] The numerous fleets thereby contributed to the phenomenon of the "tragedy of the commons"[15]: rapacious over-fishing of a resource lest other ships or nations did so first.

The transportation of fresh or frozen goods underwent rapid change on land also, thereby enabling food processors to distribute throughout the country. These changes included new railway refrigeration cars, called "reefers,"[16] and large refrigerated trucks.[17] With the expansion of the federal highway system during the 1950s, trade associations of semi-trailer manufacturers joined the US Department of Agriculture and National Bureau of Standards in examining refrigerated trailer performance in long and short runs with the goal of designing loading docks to minimize temperature changes and to keep labor costs down.[18] Still, the distribution of frozen foods remained a weak link in quality control; as late as 1966, J. Perry Lane of the Bureau of Commercial Fisheries' technological laboratory in Gloucester reported that most commercial refrigerated trucks failed to maintain proper temperatures.[19]

Another technology important to the fish stick was the supermarket itself. The first US supermarkets opened during the 1930s. Consumers flocked to them, although the Depression delayed further expansion until the postwar years, when they became a central feature of daily life.[20] During the period 1948 to 1963, large supermarket chains increased their share of the nation's grocery business from 35 percent to almost 50 percent. Managers installed high-capacity refrigerated display cases for the myriad frozen and other processed-food products.[21] In 1960, the Food Fair supermarket chain excitedly publicized the opening of its "completely mechanized, modern seafood distribution center" in Philadelphia at a cost of $100 million; it was capable of shipping more than a million pounds of frozen and fresh North Atlantic seafood weekly.[22]

Gorton's and other frozen-fish companies embraced these changes in catching, processing, refrigerating, freezing, and transporting during a time of rapid social, demographic, economic, and lifestyle changes in the United States that helped shape consumer demand and create a market for such convenience foods as the fish stick. During World War II, when women went to work in industry to replace men who had gone off to war, "convenience cooking" found an immediate market. Rosie the Riveter, the iconic woman who replaced men in industry when they went off to fight World War II against fascism, could work all day and still cook for her family at night using ready-to-eat items. A booming postwar economy and a rapidly growing population increased demand for new food products. The G.I. Bill of Rights contributed to upward mobility through educational and training opportunities for millions of demobilized soldiers. Spurred by federal subsidies and tax laws, a housing boom followed, which in turn gave rise to suburban lifestyles. Real income rose steadily and Americans spent a higher percentage of it on food. The food- and kitchen-appliance manufacturers jumped on the mass-market bandwagon to help satisfy the demand for filling these new homes. In 1952, when the population totaled 152 million, only four million families, mostly farm-based, had freezers; by 1960, this number had quadrupled. The boom in housing permitted the design of kitchens large enough to accommodate freezer and refrigerator-freezer units.[23]

In this environment, food processors sought to convert new techniques and products that grew out of military rations and meals into a "peacetime market for wartime foods," with a focus on "precooked" products.[24] To promote these new products, processors promulgated a strategy of mass marketing, new menus and recipes, and new products, of which the fish stick was one. These foods and meals built upon a tradition of scientifically based recipes in the name of public health and efficiency that originated in university home economics departments and Department of Agriculture extension services. When Quaker State Foods introduced its unfortunately named "One-Eye Eskimo" label of frozen meals in 1952, consumer response was underwhelming. Although Swanson soon thereafter made a hit with TV dinners, frozen foods were not yet a "housewife's dream." Initially, producers successfully marketed only frozen orange juice. As Laura Shapiro points out, while by 1954 the annual consumption of frozen foods had risen from seventeen to thirty-six pounds per capita, 80 percent of it was purchased by only 3 percent of the population.[25]

Still, the Birds Eye product line of General Foods introduced fish sticks to national fanfare on October 2, 1953. A newspaper article even claimed that this was "the most outstanding event" in seafood since the early 1930s. Fish sticks signaled the modern era of easy-to-prepare, nutritious foods. This shift toward precooked foods, and sea foods in particular, represented "the first big improvement in the use of raw food materials since the early days of the introduction of quick freezing." Developed at Birds Eye's seafood laboratories in Boston over a three-year period, fish sticks' time-saving quality was its greatest attribute: "No actual cooking is required," its promoters proudly announced. Just as important, the fish stick would "help increase the per capita consumption of fish."[26] Nearly simultaneously, other manufacturers such as Gorton's entered the fish stick business.

The Gorton's of Gloucester Corporation

Gorton's predecessor was established in 1906 when four fishing companies were consolidated as Gorton-Pew Fisheries Company, a firm with nearly forty vessels. Gorton-Pew grew to a thousand employees on land and an equal number at sea, with some fifty-five vessels, fifteen wharves, and thirty-five buildings. The company survived the Depression and gained business during World War II. Sales slumped after the war, however, when the demand for fish declined as meat suppliers switched from the military to the domestic market. Innovations in catching, freezing, and processing fish offered great promise only if the processors could survive the shakeout and create consumer demand. Gorton's overcame several rather lean years to become a leader in fish sticks and several other fish products. How it did so illuminates the challenges facing fish processors during the postwar years when new, scientifically tested products had to be developed, new regulatory pressures had to be met, and new foods had to be marketed in the changing retail setting of the supermarket.

In 1942, E. Robert Kinney of Pittsfield, Maine, a small town near Bangor, met Paul Jacobs of Boston. Jacobs worked for the New England Development Corporation, a Lincoln Filene–Cabot investment firm that sought to assist struggling New England industries. A history major at Bates College in Lewiston, Maine, Kinney had intended to become a high school teacher, but, preferring business to education, he left graduate school at Harvard in 1942 and moved to Bar Harbor, Maine. There he founded the North Atlantic Packing Company, a crab-canning business, and as its president and treasurer turned the struggling operation into a firm that grossed $2 million annually in 1952. He joined Gorton's in 1953 as president, where he oversaw a sevenfold increase in sales. General Mills bought Gorton's in 1968, and Kinney became that company's president in 1973.[27]

Kinney, who knew firsthand the economic challenges facing Maine's coastal fishing towns, solicited Jacobs in supporting North Atlantic Packing. Shortly after buying into the company, Jacobs became its president and the company expanded into such products as cat food made from fish by-products. North Atlantic Packing next marketed "Bar Harbor Ready-to-Fry" fishcakes in competition with Gorton-Pew. This led to legal disputes over copyright infringement on the use of "ready-to-fry," which Gorton-Pew claimed as its own. But representatives of the two companies soon recognized that it was better to combine forces; eventually, Gorton's president, McGeorge Bundy, who was later an adviser to President John Kennedy and then became head of the Ford Foundation, invited Jacobs to join the company's management.

Jacobs joined Gorton's during a difficult time: sales had declined, and 1953 marked the first time in twenty years that it had sustained a loss. (Things were not as bad as they had been in 1923, however, when Gorton-Pew Fisheries nearly went bankrupt after Benito Mussolini's government confiscated a huge cargo of cod.) Jacobs worked as director of advertising and promotion to regain appeal for the Gorton label, and he even convinced *Parents* magazine to award fish sticks its seal of approval. Perhaps because of new leadership or rekindled interest in fish products or successful advertising, by 1955, sales had shot up 27 percent. Jacobs placed salesmen in the East, Midwest, Southwest, and on the Pacific coast to market Gorton's products. While producers initially had trouble convincing supermarkets to expand freezer display-case capacity to carry frozen products, Kinney and others eventually succeeded in placing fish sticks and other ready-to-serve items in such grocery outlets as A&P, D'Agostino, Diamond K, Food Fair, and Star.

Like other producers, Gorton's devoted great attention to advertising, packaging, and display to generate consumer interest in fish sticks. The advertising firm of Petley, Clark, and Johnston investigated Birds Eye's fish stick advertising and in October 1953 recommended that Gorton's embark on a national campaign; the firm's president, G. E. Petley, sent Jacobs a series of newspaper and magazine clippings to underscore the potential of a similar advertising campaign.[28] Welles Sellew, a Gorton's vice president, followed with a memorandum to all brokers on "heat and eat" seafood, especially fish sticks, pointing out their convenience, quality, and great taste: "This is a new item— the potential for this product is *terrific*."[29] Manufacturers of fish sticks stressed their wholesomeness, modernity, and time-saving qualities: the "harried housewife" could "heat and eat" fish sticks, which were nutritious, based on scientific standards, and used only such wholesome ingredients as potatoes, salt cod, shortening, and meal. Jacobs likened the success of fish sticks for the fishing industry to that of "frozen juice . . . to the citrus industry." There was "seemingly no limit to the potential market" for fish sticks.[30]

Researchers, engineers, package designers, and salesmen were involved in the development of this new product. Jacobs glorified it in his standard stump speech, "The Fabulous Fish Stick Story," which he delivered during the early 1950s to business, community, and other groups. He called the fish stick a "tribute to the ingenuity of the American business man" and praised the product's uniformity and simplicity. The consumer had only to open the package, heat the contents, and serve. While cod and canning had a long tradition in New England, canned fish had never duplicated the flavor and appeal of fresh fish. Birds Eye's quick freezing allowed frozen fillets to be shipped anywhere and maintain their fresh taste. Jacobs observed that the fish industry thrived during the Depression because fish products were economical in cost, and it continued to grow during World War II when meat was in short supply, but as meat rebounded during the postwar period, it threatened to win the "battle of proteins." Modern boats, electronic location devices, better, more efficient trawling equipment, and new methods to cut and process, skin, and fillet permitted increased output, but these could not create consumer demand.[31] That job was up to the fish-product industry.

To engender "a positive response in most people's minds," Jacobs promoted an extensive advertising campaign to educate consumers about how fish had as much protein as meat and was easier to digest. He admitted the challenges involved, since "most women do not like to handle fish [and] most women don't know how to cook fish properly." Birds Eye, Fulham Brothers, and Gorton's— all Massachusetts companies—introduced fish sticks in the spring of 1953 as the answer to those challenges. They had solved the problem of ending up with dry, tasteless fish by undercooking fish sticks at the factory to prevent overcooking at home. Simple instructions saved the day, as did an appealing package: "Here I am—I'm a Fish Stick—take me home and try me out—I taste good!"[32]

Gorton's rose to the top among competitors, and it expanded its facilities and products rapidly. It built its own $1 million processing plant in Gloucester, which it opened in 1956 as the Gorton's Seafood Center. Over the next ten years Gorton's acquired other companies and also entered international markets. Its other facilities serviced shrimp, fish by-products, and institutional sales, all of which it offered through the acquisition of such other businesses as Florida Frozen Food Processors, Canapro of Canada, and Blue Water Seafoods, a subsidiary of Fishery Products. In addition to its fish sticks that used less filler and bread than other fish sticks and its "Bake 'n' Serve" dinners such as fish steaks in cheese sauce, the company now marketed scallops, perchies (breaded and cooked perch fillets), and other dishes in "family-size" packaging. The company expanded sales to restaurants, hotels, caterers, and schools, and it vigorously pursued the manufacture of easy-to-serve "portion-controlled

Table 1   Gorton's of Gloucester major products, 1906 and 1957

1906	1957
Codflakes in glass jars	Main courses: sole, flounder, and other fillets
Water Lily brand boneless codfish	
Perfect, hand-picked codfish	Fried/frozen specialties: fish sticks, cakes, and perchies
Codfish cakes (boneless)	
Smoked bloaters	Frozen fillets: perch, sole, codfish, whiting, pike
Boneless herring in glass jars	
Sliced, smoked halibut in glass jars	Frozen, portion-controlled seafood for institutions: "uni-fill-etts," fish sticks, patties, cakes, and others
	Salted and pickled fish by-products

items." The impact on Gorton's of innovations in refrigeration, freezing, and processing technologies and of social-demographic change on the product mix between 1906 and 1957 is indicated in table 1.

The table also reflects how its inventory changed from salted and cured products that required overnight soaking to remove excess salt, to those that a housewife could simply pop into the oven with little mess or bother. Other innovations followed during the coming years, such as the four shrimp products and six new lines of fish fry that were added in the early 1960s. But the most important items for Gorton's postwar rejuvenation were fish sticks and portions.

Gorton's brochures touted fish sticks as "the industry's greatest contribution to modern living." "Thanks to fish sticks," one brochure read, "the average American homemaker no longer considers serving fish a drudgery" but a pleasure because they were easy to prepare, "thrifty to serve and delicious to eat." Fish sticks "have greatly increased the demand for fish while revolutionizing the fishing industry." To accomplish this, first solid blocks of clean, white, frozen-fresh fillets were cut into stick sizes with as little waste as possible, using modern band saws. Because processors preferred thinner band-saw blades, which limited the waste of food substance during cutting, there was an effort to improve the teeth, hardness, and welding of the blade ends. There was, however, a fine line between the life of the saw blade and its thinness, as the thinner the blade, the faster the teeth wore down and the more often the band broke. When this happened production came to a halt, and lost production time was more costly than the blade itself. After cutting, the fish sticks were covered with "just the right percentage of breading for maximum taste appeal."

Next, conveyors conducted the fish through automatic fryers, where they acquired an "appetizing light golden-brown hue" in fat heated to between 375 and 400 degrees Fahrenheit. From there, it was on to packaging, labeling, freezing, and shipment.[33]

Once at the supermarket, Gorton's, Birds Eye, and other manufacturers encountered problems related to freezer display cases, including significant temperature variations. Too often, store personnel placed frozen products above the load-line in display cases so that products at the top front and middle of the cases were stored at higher temperatures than those at the bottom and rear. While the latter were acceptable insofar as product maintenance was concerned, they were unacceptable regarding display, marketing, and sales. And, of course, temperature variation could lead to rapid deterioration in quality. Studies showed that frozen products encountered their highest temperatures in local retail deliveries, that is, just before the products passed to the consumer. These problems interfered with the creation of a stable niche for such frozen foods as fish sticks in the market.[34] In addition, Gorton's salesmen competed with those of other companies to secure freezer display space in supermarkets and to instruct supermarket personnel on their products' proper handling and display. Gorton's instructed its representatives to always keep frozen-food displays "neat and attractive," and it gave them a brochure explaining not only how best to do this, but even how to park: far from the front doors so as to leave the "close in, important spaces" to customers. Salesmen were to talk "with the manager and stocking clerk together if possible, be enthusiastic, reveal charts and pictures showing how to arrange foods." When suggesting how to arrange Gorton's products in the cases, they were told to watch for blind spots, but to be fair to other companies' products in arranging the case. "Show the manager what you've done and why you have done it. Show them sales statistics" demonstrating how volume will grow rapidly.[35]

As noted, fish sticks met with immediate success. Within months of their introduction in 1953, they had grabbed 10 percent of noncanned fish sales. Production leapt upward; in Gloucester alone, 500 new jobs were created and processing went from a seven-months-a-year industry to year-round. According to the *Wall Street Journal*, fish sticks were "the first really new processing development in a quarter-century for one of the nation's oldest industries." In 1953, thirteen companies produced seven-and-a-half million pounds; in the first quarter of 1954, another nine million pounds were produced and fifty-five new companies entered the business. Fulham Brothers of Boston expanded into fourteen cities, with a resulting tenfold increase in sales from September 1953. A sampling program in Detroit that involved "hitting 200,000 families with coupons" proved successful. While fish sticks were slightly more expensive than frozen fillets (which were neither breaded nor cooked), company representatives claimed that their breading saved the housewife both time and money.[36] In 1954, supermarkets reported a 30 percent sales increase of fish sticks over the previous year, and Gorton's was running one hundred workers in two shifts and expanding facilities wherever it could. As the corporate newsletter, "The Man at the Wheel," commented: "Acceptance of fish sticks . . . has been so widespread that many expect them to do for the fisheries what fruit juices have done for the fruit trade."[37] Their quality, attractive wrapping, and advertising had been the key to success.

Gorton's instructed salesmen that women were the shoppers and buyers and that in marketing fish sticks, they needed to approach men and women differently. According to a survey whose origin and conception is unclear, of a thousand individuals randomly selected from three million subscribers to *Ladies' Home Journal* in autumn 1954, almost two-thirds of the 738 respondents knew about fish sticks, and most said they were available at local stores. Yet only one-third was familiar with specific brands—10 percent with Birds Eye, 7 percent with Mrs. Paul's, and 4 percent with Gorton's—and fewer than 40 percent had served the product.[38] A Gorton's internal memorandum (with the apparent goal of increasing name recognition among housewives) used simple drawings to indicate the influence of the female and male subconscious on purchasing behavior. When women heard the word "cheesecake," the memo indicated, they thought of food; on hearing the same word, men thought of a sexy woman. So the question was, how to appeal to 55.5 million women in the United States, 90 percent of whom were homemakers—meaning cook, driver, mother, nurse, host, teacher, and "purchasing agent"—and fourteen million of whom also worked outside the home? Gorton's 1958 sales promotions therefore aimed at the busy women who bought the groceries. Its "Busy" campaign adopted the slogan, "Let Gorton's Famous Seafood Chefs Cook for You." Advertisements in women's magazines for its "Heat 'n' Eat Seafoods"— fillets of sole, flounder, steaks, scallop casseroles, and fish sticks and fishcakes in various sauces, cheese, or breaded—focused on this market.[39] By 1960, Gorton's billboards had spread from New England to Florida,[40] and fish sticks with them across the nation.

Federal Regulation, Federal Funding, and the Fish Stick

Government regulatory and purchasing programs significantly helped the fish stick. Federal efforts to regulate frozen seafood accelerated after World War II; in fact, culminating with the Food Additive Amendment of 1958, the government had become active in ensuring a healthful and safe food (and drug) supply. The regulatory history of various industries, processes, and labor practices for public safety is not the focus of this book; suffice it to note that leading manufacturers and businesses, while often initially opposing regulation, have eventually endorsed it as a way to reduce cutthroat competition. In virtually every aspect of the food industry, self-proclaimed ethical manufacturers welcomed regulation, while cut-rate producers feared it because it would mean they would face higher costs in order to meet higher standards. Regarding fish sticks, many producers initially opposed regulation as interference in commerce, yet in the face of daunting competition among an increasing number of producers and a decline in product quality that shook consumer confidence, the leading suppliers eventually embraced it.[41]

Evidence had accumulated that the frozen-fish industry could not guarantee quality without some kind of inspection program in place. Francis Schuler, an economist with the Bureau of Commercial Fisheries, concluded that lower costs in an increasingly competitive industry had been attained only by using shortcuts, leading to loss of quality of the final product. In a series of articles during the 1950s and 1960s, *Consumer Reports* noted a general decline in the quality of fish sticks as more producers entered the market.[42] When consumer dissatisfaction caused a decline in production in the mid-1950s, firms such as Gorton's called for voluntary standards of government inspection of fish sticks. Believing that a US Grade A inspection seal on packaged products would "erase many of the suspicions the public had toward all frozen seafoods," Gorton's marketing people decided to embrace standards as a way of separating their fish sticks from those of competitors. Consequently, the Department of Agriculture published US Standards for Grades of Frozen Fried Fish Sticks (effective August 21, 1956, re-codified July 1, 1958) as recommended by the Bureau of Commercial Fisheries of the US Department of the Interior. This voluntary program stipulated standards, inspections, and certifications for sticks and portions.[43] Grade A fish sticks had "good flavor and odor" and good appearance and were practically free from defects—that is, with no identifiable bones, no broken or damaged sticks, and no blemishes. Leading manufacturers promptly implemented the standards, even though enforcement increased production costs, since it required the funding of continuous inspection of fisheries' products by full-time federal inspectors at all plants. By December 1958, sixteen fish-processing plants, located primarily in the Northeast and Gulf states, had adopted the continuous-inspection standard. By March 1960, forty-one inspectors had placed thirty-two plants in the industry under continuous inspection, and some 92 million pounds of fish passed under their noses in that year alone.[44]

Inspectors turned up a series of problems that illustrate the significance of inspection. One inspector had rejected a certain lot of frozen fish, only to learn that the same lot turned up at another plant, a hundred miles distant, where another inspector also rejected it.[45] Yet if fish sticks were to survive as a product, more than voluntary inspection standards were necessary. In February 1961, the editors of *Consumer Reports* published a highly critical analysis of the product. Of twenty-six brands (312 samples), only two received "acceptable" ratings. The others were deficient "apparently by poor storage and handling practices." Worse still, fully 25 percent of the sticks tested did not meet the requirement that a stick was to be "clean," "wholesome," rectangular, and containing not less than 60 percent fish flesh by weight.[46] Were these fish sticks or breadsticks?

By 1967, the National Fisheries Institute supported a bill that provided for mandatory inspection of fish and fishery products.[47] "Let's face it," the editors of *Quick Frozen Foods* insisted, "whether it is a mandatory inspection set by the industry or by the government, *the ultimate judge of our products [sic] quality is the consumer herself.*"[48] The Wholesome Fish and Fishery Products Act of 1968 provided, after a six-year grace period, for mandatory and continuous inspection standards for quality and cleanliness and the approval of both labels and packaging.[49]

E. Robert Kinney, Gorton's president, defended the government's standards. In frozen seafood, "as with so many other new and expanding industries," he wrote, "we are prone to believe that recently proposed governmental regulations will needlessly complicate our lives." But industry "must learn to live with [the laws] gracefully, and take advantage of what they can offer us in consumer confidence and acceptance of our products." He claimed that Gorton's had pioneered continuous government inspection of seafood with the opening of its new Seafood Center in 1956. Gorton's considered inspection "an important facet of the great American search for constantly improved productions to meet new and changing appetites." Inspection and advertising that touted goodness and wholesomeness were the keys to developing consumer confidence in new convenience foods such as breaded shrimp, breaded portions, and fish sticks.[50]

Several other federal programs also helped improve sales of the fish stick. One was the Saltonstall-Kennedy Act, passed by Congress in 1956, that supported fisheries research and provided $45 million to promote the virtues of new products in supermarkets, including the fish stick. Senators Leverett Saltonstall and John F. Kennedy, Republican and Democrat of Massachusetts, respectively, introduced this legislation for several reasons: to support commercial fisheries during a recession in the industry that was caused in part by growing foreign competition; to convince a skeptical public of the value of fish sticks and other products; and to encourage supermarkets to undertake the large expense of purchasing and installing freezer display cases.[51]

The federal government also supported the fish stick through school lunch programs, and suppliers took advantage of this. Building on programs run by charities, federal and state governments had expanded school lunch programs during the Depression in part to provide a market for excess farm products.[52] In 1946, Congress, stressing military expediency, passed the National School Lunch Act "as a measure of national security, to safeguard the health and well-being of the nation's children and to encourage the domestic consumption of nutritious agricultural commodities" through the establishment, maintenance, operation, and expansion of nonprofit school lunch programs. The number of children, number of meals served, and amounts of food used all increased rapidly, as there seemed to be universal recognition of the act's benefits. In 1944, nearly 3.8 million children participated, which swelled to more than six million by 1947. By 2004, federal funding for 27 million school lunches daily, together with food-assistance programs to mothers and small children and several other programs, cost approximately $16 billion.[53]

Fish stick managers found a ready-made market in school lunches. The inexpensive yet wholesome fish stick, served with potato chips, orange juice, salad, rolls and butter, cake, and milk, would be a "favorite school lunch entrée." Gorton's first tested fish sticks in Manchester, Massachusetts, public schools. They had to be served "piping hot," which left great responsibility in the hands of the food-preparation staff. The company was confident this would not be a problem and suggested a variety of options to keep fish sticks as a high-demand item in schools: they could be served as fish burgers with tartar sauce on a bun; as hot appetizers; with scrambled eggs at breakfast; and so on.[54] Gorton's worked tirelessly during the 1950s to promote fish sticks through distributors to school lunch programs,[55] and to this day they maintain a spot in many school cafeterias.

Fish Sticks and Professors

One last input was crucial to the success of the fish stick: scientific research and development in various forms. In 1953 or 1954, Gorton's management commissioned a detailed report by Harvard University economists to determine the proper long-term strategy to secure their position within the industry. The marketing specialists noted that in spite of a postwar decline in demand for fish that had hurt Gorton's and other suppliers, fish sticks were ready to hit the market. Surveys of producers, distributors, and grocers indicated that fish sticks were "extremely easy to handle, the housewife simply heating them up in the oven." While New York's Fishery Council for fresh fish raised the concern that fish sticks "may turn out after all only a 'sizeable flash in the pan,'" the Harvard report indicated that greater faith in the future of fish sticks was warranted. The management specialists called for extensive promotion of distinctive Gorton's packaging to attract consumers and replacement of the bulk pack with family-size packages to keep Gorton's "out of the price-cutting mill race." They also recommended that Gorton's quickly modernize facilities to include a continuous frying facility like the one recently built by Birds Eye.[56]

Food science and technology also supported the fish stick.[57] To give just one example, in 1956, Paul Jacobs asked researchers at MIT's Department of Food Technology to help Gorton's improve fish sticks, especially in the areas of quality control and efficiency of production. The department could determine when a product was best, how to produce it in the most efficient way, and how to invent a new product that was novel and appealing in taste and convenience.[58] To promote closer ties between research and food, between science and industry, in July 1956 Gorton's and MIT sponsored what the organizers claimed was "the first seminar in the history of the frozen foods industry." The seminar focused on "how to keep frozen seafood sufficiently frozen at all times from plant to customer to maintain flavor, texture, nutrition and kindred values inherent in wholesome fresh fish." The journal *Food Engineering* called the seminar a "unique [and] trailblazing researcher–marketer" session. The seminar's participants visited several supermarkets and the automatic seafood-freezing plant recently opened by Gorton's. Professor Bernard Proctor, head of MIT's food technology department, committed his department to a long-term effort to assist Gorton's in improving the fish stick.[59]

Because of its proximity to Gorton's and its long tradition of involvement in food science, MIT was a logical choice to help improve the fish stick. From its founding in the late nineteenth century, MIT's scientists had conducted research on "industrial biology" in the service of the food industry, including efforts to preserve and package products for the canning industry.[60] Its extensive efforts in food preservation continued through 1944, when the independent Department of Food Technology was established, whose faculty worked closely with a number of food-processing firms that were investigating ways to apply new technologies to the food-service industry. Two MIT students even wrote their theses on the topic of fish sticks.[61]

MIT scientists joined Gorton's directly and indirectly in product improvement: concerning direct involvement, they developed the "fresh-lock" process, which Gorton's marketers claimed prevented loss of "natural fish flavors, texture, color and nutrition"[62]; indirectly, MIT scientists conducted research on "irradiation pasteurization" to prolong the shelf life of seafood and other foods. MIT received funds for this research from the food and container industries, the Atomic Energy Commission (AEC), the US Army Quartermaster Corps, and the Public Health Service. In 1966, AEC scientists petitioned the US Food and Drug Administration to allow commercial use of irradiation on cod, haddock, pollock, flounder, and sole, having concluded that shipboard irradiators were feasible. One such device was installed on the ship *MV Delaware* to test radiation preservation at sea. Ultimately, however, it was found that radiation adversely affected the taste, texture, color, and shelf life of fish products, no matter how they were packaged or handled; proper refrigeration and sterile facilities were just as crucial and ultimately cheaper.[63]

The Fish Stick on Your Table

The fifth anniversary of Gorton's Seafood Center on August 25, 1961, celebrated the close ties among technology, science, and fish sticks. The event drew 250 visitors, guests, and speakers from Gorton's, government, science, and labor—all involved in the evolution of fish sticks, the participants behind the packaged product. Senator Benjamin Smith II (D-MA) praised the connections between federal subsidies and the fishing industry, declaring that "the government's revenue and its success depends [*sic*] on business profit and business success." Smith, whose father was a former president of Gorton's, added: "I hope that the foresight and vision of this company will be contagious not only in industry but also in the community."[64] Samuel Goldblith, deputy head of MIT's Department of Food Technology, presented his speech, "Industrial Research Is the Answer to the Challenge of Ignorance about Seafoods." The ignorance he alluded to was the consumer's reluctance to buy fish products, and his presence reflected his mission to overcome this ignorance and promote the diffusion of irradiation technologies throughout the fishing industry. Goldblith, a director of Gorton's since 1959 and consultant to other food companies, stated that "[t]he key to leadership in this industry is scientific research and a gamble of time, people and money." After this speech, Kinney presented the professor with a check to support additional research on seafood freshness by an MIT graduate student.[65]

While frozen seafood sales generally increased during the 1960s, fish stick sales were uneven, in part because competitors were bringing lower-quality products into the market. The number of frozen seafood processors had increased by 774 percent in more than twenty years, from 107 in 1947 to 935 in 1967, 529 of whom produced frozen fish products and 406 that produced frozen shellfish. There were 120 cod processors, 89 sole, 81 haddock, 74 salmon, and 34 fish stick processors,[66] and standards went only so far in maintaining quality. Another reason was that several producers decided to focus on the more lucrative "portions" commercial market, as opposed to the smaller retail market for fish sticks; further, a papal decree ending "fish Fridays" for Catholics seems to have contributed to a decline in demand, especially in school lunch programs. By 1962, breaded fish portions had surpassed fish sticks in production at 78 million pounds, a 30 percent increase over 1961. Portions—uniformly cut two-, three-, or four-ounce segments of ground fish—generated sales, especially among such institutional users as restaurants and hospitals, because of their consistent size, ease of preparation, and cost effectiveness.[67] Institutional sales of portions, shrimp, and breaded shrimp dominated processors' sales. And as for the retail market, fish sticks—at some 80–100 million pounds and $70–80 million in sales annually through the 1960s—fell to second, third, or fourth, depending on the year, behind fillets, portions, and shrimp. Yet, while not a major seafood nor an institutional favorite, the fish stick was here to stay.[68]

Like many processed foods, the fish stick occupies an important place in the American diet and the American marketplace. A visit to any neighborhood supermarket reveals fish sticks produced by no fewer than a half-dozen companies. Still, it is not a staple of the diet. To most Americans, fast food means beef, not fish, and even when fish is chosen it is more likely to be breaded fillets than fish sticks. Among factors that prevented the fish stick from becoming as celebrated as the hamburger were the American palate, which often suffers from landlubbers' taste buds, and the declining quality of many fish stick products after their initial introduction. As more and more companies entered the market, many of them sought to cut costs by producing an inferior product: one with more breading and other nonfish substances. The fish stick became the "hot dog" of the ocean; many Americans continued to feel that fish sticks were a mediocre food, and companies that strove to keep quality (and fish) at the forefront of production suffered the consequences. This may have been because producers created demand for a product after fortuitous technological innovations and social pressures had combined to create the fish stick. Ultimately, the "piscatorial engineer"—part advertiser, part salesman, and part food-product innovator—could only put fish sticks on the table, but couldn't make consumers fill their plates.

Jogbra: No man-made sporting bra can touch it (advertisement late 1980s, with two of Jogbra's co-founders as models, Hinda Schreiber and Lisa Lindahl). Demand among female athletes and other active women for more comfortable sportswear with greater support drove the development of the sports bra. Courtesy of Hinda Miller and Archive of the Smithsonian Museum of American History.

2

The Sports Bra

Gender and Technology

In 1967 Kathrine Switzer completed the Boston Marathon—against the wishes of Jock Semple and the Boston Athletic Association. In 1972 Title IX of the Educational Amendments banned sex discrimination in schools, including unequal athletic opportunities. In 1984 Joan Benoit won the first women's Olympic Marathon championship in Los Angeles. And, barely noticed at the time, in 1977, two graduate students introduced the first sports bra fashioned from two jockstraps. Today's sports bra is the result of thirty years of study of physiology and biomechanics, developments in material science, changes in fashion, impetus given to participation in sports by Title IX, and notable achievements among woman athletes—Kathrine Switzer, Joan Benoit, Brandi Chastain, and many others.

For many people, the iconic moment for the sports bra occurred in 1999 when Chastain scored the winning penalty kick that led the US women to the World Cup in soccer and then removed her uniform top in celebration, revealing her sports bra. Twenty years earlier Lisa Lindahl was a 28-year-old graduate student at the University of Vermont working as a secretary. Also a serious runner, she desired more comfort and support in a running bra that would not bind or chafe. She approached a childhood friend, costume designer Polly Smith, and Hinda Miller. Miller graduated from New York University and had a position in the autumn as an associate professor at the University of South Carolina; she worked that summer for the University of Vermont as an assistant costume designer at the Lake Champlain Shakespeare Festival, backstage with Polly Smith, her boss and fellow NYU graduate, who introduced her to

Lindahl. During their free time the latter two would go out running. While enjoying the endorphin rush, they both experienced discomfort. Miller wrote, "there was a downside: our breasts jiggled uncomfortably when we ran. Our nipples became sore from chafing against our sweaty T-shirts."[1] The three women determined to design a bra for running and other activities. They worked their way through the local bra shop to see if something in stock might be modified to meet anti-bounce, anti-chafing, comfortable-fit criteria.[2]

While various support, protective, and enhancing devices for the female breast have existed for hundreds of years, the modern bra, judging by patents and marketing information, dates to the late nineteenth and early twentieth centuries. By the 1920s Maidenform became a leading manufacturer; other companies joined the market and learned to build product loyalty with younger women. Bras evolved to emphasize curves and breast size, not to hide the breast, nor necessarily to provide adequate support and comfort. Directly and indirectly, military developments influenced further advancements in bra designs. Among many things required for the World War II effort (metals, agricultural products, and so on) were natural fabrics for uniforms and the like. This pushed the creation of synthetic fibers to meet natural fabrics shortfalls that were adopted in the apparel industries. These fibers—rayon, acetate, nylon, and Lycra—improve comfort, fit, and durability of many items of clothing, including bras.[3] Sports bras were built on new materials and marketing, but also required a big market. That came to fruition throughout the following decades, when increasing numbers of women turned to sports and exercise.

Lindahl, Smith, and Miller were pondering the sports bra when Lisa's husband walked into the room and jokingly pulled a jockstrap over his head and around his chest. Intrigued and amused, they bought two jockstraps at a local bookstore and sewed them together. The device offered "bra" compression, straps, and support across the chest wall and the back. These young designers put all seams on the outside to prevent discomfort or bleeding. Polly assembled the first Jogbra from the jockstraps, while Lisa tried out the design at the track, with Hinda running backwards in front of her to judge its function. It worked. The designers were enthused and empowered by their success and determined to market the sports bra, although with little business knowledge about how to embark upon such an undertaking.[4]

As Miller pointed out, the sports bra was form following function, unlike the first bra, which had been a male fantasy of the female body with form coming before function. In the case of the sports bra, while function came first, materials more appropriate for exercise, moisture removal, and support came along soon thereafter: these new materials made each version of the sports bra more functional. When polycotton (a polyester-cotton mix) Lycra became available, the possibilities seemed limitless: cotton provided comfort, poly offered strength, and Lycra contributed support.[5] In a way, as LaJean Lawson recalled, the first sports bra was an "anti-technology"—a technology derived from a function for another gender, and an old technology at that. But the two jockstraps provided a wide bottom band so crucial in sports bra designs; it was compressive; used no metal fasteners; and with the leg straps crossed in the back it did not permit slipping. The sports bra provided solutions for support, straps, and anti-chafing in one design.[6] Many historians have explored how science and technology privilege first the male and the masculine and make the woman the invisible "other." In this case, perhaps, science and technology combined to assist women athletes, dancers, and exercise aficionados to occupy a major public position in the sports world with vastly increased numbers of participants.[7]

I do not pretend to provide a full treatment of issues of science, technology, and gender in this chapter, but rather I introduce ideas about the ways in which such concerns as class, race—and in this case gender—shape technologies and in turn are shaped by technology.[8] Some technologies, like birth control devices, have been produced specifically for women.[9] Others, like kitchen appliances, are associated with women. In *More Work for Mother*, Ruth Cowan shows how the evolution of such home labor-saving devices as the stove and microwave did not necessarily give women free time for other pursuits, but were accompanied by changing and greatly increased ideals of cleanliness and hygiene. While women's labor may have been less arduous, they faced higher standards and hence "more work."[10] In a study of laundry work and the laundry business between 1880 and 1940, Arwen Mohun examined how the development of washing machines joined with new attitudes about cleanliness to change the way women did laundry. At the same time, businessmen opened commercial laundry facilities where primarily women laborers toiled and they faced hostility in any attempt to unionize both from bosses and from the American Federation of Labor, largely because of their gender. When the electric washing machine entered into the middle-class home, in the United States African American women began to replace white women as laundry workers. At the same time, the laundry businesses declined as they could not compete with home washing—where labor was also "free."[11] Kitchen design and the role of the female "homemaker" reflect the belief that women, as the foundation of the family, should be responsible for keeping the house clean, working efficiently, and not wasting time on out-of-the-home financial or intellectual pursuits.[12]

In the sciences, women have long been excluded from being equals as researchers, professors, and department chairs, at the dawn of modern science were considered the weaker sex physically and intellectually, and were forced to participate in the evolution of science outside of academies of sciences and universities.[13] From the early nineteenth century until quite recently, many scientists believed that women's brains were smaller, or incapable of achieving the stellar contributions of men in the "hard sciences" like mathematics and physics.[14] In other fields, for example, primatology, males applied their allegedly objective understandings of race and class—in fact, their attitudes of the superiority of white civilization and males—to the study of apes to apply bourgeois notions of the nuclear family to chimpanzees and gorillas.[15] And to this day women continue to face significant obstacles to entering the work force as equals of men and, in many countries, including the United States, and including the sciences, are still paid significantly less than men for the same work.[16] Fortunately for female athletes, women entrepreneurs, scientists, and others understood that to wait for technological change to come from male-dominated notions of athletics and attire would be fruitless, and women specialists quickly achieved novelty and success in sports equipment.

Pioneers in Sports Equipment

LaJean Lawson is probably the only person with two advanced degrees in "sports bra science," an MS in clothing and textiles and a PhD in Exercise Science. She has conducted biomechanical and wear-testing research projects since the mid-1980s. She left academia to work with Champion Athletic Wear; she reasonably allows that she is "probably the 'grandmother' of the sports bra industry." Lawson comes at sports bras not only from the scientific point of view, but from the social point of view. That is, science for Lawson is not an outcome in itself, for example, empirical testing and publications, but makes life—in this case the life of women athletes—better through better understandings of their activities, desires, physiology, and biomechanics. A sports bra must above all else feel good. This is a difficult task given that topology, ground forces, women of different shapes and sizes, and hot and sweaty repetitive exercises lead to abrasions and bruising, which must all be managed.[17]

In the early years of sports bra development, the designers fought for comfort and stability. Jogging and aerobics had become crazes, but many women felt uncomfortable because of breast motion, or as they called it "flopping about" and "boob-bouncing." Bulky, more traditional bras were simply inappropriate because of the energy and impact of dance aerobics and because women did not want what they considered to be ugly bras under their neon, Lycra tank tops and leotards. Lingerie companies pursued sports bras, sensing a new market in search of a more appealing look, and they also sought to design comfortable bras. But they discovered there was much more to sports bras than colors and fabrics. They needed to understand biomechanics—how breasts move in motion—and materials.

In the first bras, designers sought simplicity in design; they knew that minimizing bounce was more realistic than eliminating it. Lindahl, Smith, and Miller worked at home to try out a few prototypes, drawing on their unscientific knowledge of sport and their experience in costume design for the theater.[18] Miller's father lent them $5,000, and the brand-new company worked with an equally nascent apparel manufacturer in South Carolina to produce the first marketable Jogbras. The bras were distributed primarily through mail order in 1978. By the following year Lindahl and several newly hired product representatives were targeting sporting goods retailers.[19] They refined the Jogbra and went to market creating Jogbra Sportsbras.

In the autumn Miller moved to South Carolina to teach, with $5,000 loaned to the endeavor from her family, and stumbled into Carolyn Morris, a recently unemployed seamstress looking to start her own business, who began to produce the Jogbras. She quickly manufactured forty dozen "jock-bras." They realized quickly the need to change the name, and the Jogbra was born. In 1978–79, Lisa sold 240 bras in Burlington and San Francisco, Hinda an equal number in New York, North Carolina, and South Carolina. The packaging was a transparent ziplock bag with "Jogbra" printed on the diagonal, in sizes S(mall), M(edium), and L(arge). The business boomed.[20]

Yet selling sporting goods stores on the Jogbra was difficult because of their novelty and because many store owners were male, ignorant of the need, did not appreciate the innovation, and probably worried about giving up floor space to ancillary equipment. The masculinized construction of gender in the sporting world to this day remains pronounced in the types and levels of endorsements for male versus female athletes and in the type of sports that are characterized as women's sports (gymnastics, figure skating, beach volleyball) as opposed to "men's sports."[21] Yet personnel in sporting goods stores have overcome their ignorance about women athletes, their activities and equipment, not the least because of the motivation to sell equipment to larger numbers of customers in such a lucrative market. At the early stage of the development of the sports bra, although Title IX had become law, the numbers of women athletes had not reached a sufficiently critical number to justify setting aside store space for sports bras.[22] The Jogbra had only three sizes, but this was enough at the early stage of sales, precisely because the important task was to place the product in sporting goods stores. The young men who tended to manage and stock those stores felt confusion or lack of understanding about the sports bra as it was; they truly did not understand women athletes and women's biomechanics. But Lindahl and Miller were determined to go to "mom and pop stores," and to push the item not as a bra but as a piece of equipment. They had to educate store managers and convince them to treat the sports bra just like other stock items.[23] This was, Miller says, "virgin territory and we were a small company." Yet they pushed and prodded to make headway. "We got more sophisticated and went to more sizes to get into department stores. And once Playtex got into it they had no trouble filling stores."[24]

Hinda Miller recounted how they "wanted as much function as possible." Yet they wanted something attractive as well because women wanted to take off their shirts, too. They added spring, summer, and autumn colors.[25] Quickly they developed sports tops because many women did not want their abdomens showing. Indeed, the designers understood that each body was different. They also had an important belief that no matter the woman's age or shape, she had the right to the benefit of exercise. They determined to design a variety of sports bras for this. They worked with Lawson, whose graduate research had led to motion control requirement charts in which they attempted to match body type and structure with level of activity. The overarching goals were safety, feeling good, and comfort.[26] By the time Miller and Lindahl sold Jogbra to Playtex Apparel in 1990 (which was quickly bought by Sara Lee, whose Wonderbra was the rage), the company employed more than 175 people and had $75 million in annual sales.[27] Miller stayed on as division president, overseeing a transition to the Champion brand.[28]

Other companies recognized the need to enter the market as more and more women exercised and joined organized sports. A Nike spokeswoman observed, "The relationship [between the Playtex Jogbra and Nike's sports bras] is one of pure inspiration—clearly this first sports bra by Playtex revolution-ized how people thought about bras for athletes and it inspired a movement of research and innovation behind what we believe is the most important piece of equipment for female athletes."[29] Nike introduced its first line of women's ap-parel in fall 1979 and included sports tops. Its three-quarter length Airborne silhouette debuted in the early 1990s and provided basic compression shelf lining (two layers of material that compressed the breast tissue to the chest wall). Nike undertook their first preliminary research on breast motion of fe-male athletes in the mid-1990s at the Nike Sports Research Lab. These in-sights led to the evolution of its first shelf/compression sports bra in 1999, and to a bonded, no-sew sports bra in 2006 called Nike Revolutionary Support that introduced encapsulation and adjustability. According to *Advertising Age*, it improved comfort and reduced vertical motion by 30 percent as compared to Sara Lee's Champion Action Tech bra.[30]

Support, Comfort, Design

An accomplished marathoner (now approaching her fortieth 26-mile finish) writes of the difficulties in finding appropriate attire in the 1990s: "I used to be 50+ pounds heavier than my current weight and I can certainly appreciate a high-quality sports bra. When I first started running in 1996, I went through many test bras before I could find one that did not chafe, cause pain, or made me jiggle too much. I returned many sweaty bras to the poor cashiers at Road Runner Sports." Even when she found the right encapsulated bra, she still had to "line myself with band-aids along the bra band" to prevent blistering and also had to use a lot of petroleum jelly. She notes, "Losing the extra weight helped because I can now wear compression bras without the more abrasive and binding elastic used for the bra band."[31]

Obviously, there's no "one size fits all" for the sports bra or for the activities that athletes pursue. Athletes must choose carefully among several different manufacturers, materials, and designs. Generally, there are two approaches to the sports bra. One is compression, to bring the breast as close as possible to the chest wall and restrict its movement. The other is encapsulation, where the structure meets the size and shape of the breast. Some women refer to the latter as the "lift and separate" bra. There's also a cross between encapsulation and compression in which bra designers attempt to balance size and the nature of the breast (loose or pendulous tissue, for example). Good fit, good coverage, and good materials are the keys to achieving four goals: minimize breast motion, reduce friction, increase modesty coverage, and use lines and colors for a bit of personal style.[32]

Support comes from four main points on a sports bra: the shoulder straps, the cups, the chest band, and the wings of the bra. Compression styles are designed for medium-impact activities and work best for smaller cup sizes; women with larger cup sizes may prefer individual encapsulated cups in a bra with more adjustability. The racer-back design provides increased range of motion through the back and shoulders and provides the basis for a range of styles; it anchors the bra to the body, preventing the straps from sliding off the shoulders. Sports bras with adjustable straps and chest band are often preferred by women with larger chest sizes because of their ease of use and high support. The chest band contributes to stability—for larger breasts the key is additional under-bust support.[33] Yet other women reject adjustable straps because of the need continually to readjust them. Women typically prefer a tank or long bra style with a shelf bra for light- to medium-impact activities.

But it is all ultimately about understanding the garment-body interface, the way shoulders rotate during exercise, and how the bra tugs hundreds of times, which requires lower friction materials. The design must recognize the human skeleton and be based on body mapping, especially for a bra used in such a long event as a marathon. Manufacturers are still learning that they cannot embrace every new technology, process, or fabric simply because it is innovative. Nike introduced a bra some years ago with a glued seam that turned out to be itchy. Tacking and gluing did not work, the silhouette was unattractive, the bra did not support properly, many individuals disliked it, the model was discontinued, and Nike returned to the sown seam with a soft yarn. As Lawson pointed out, if you use technology for the sake of technology, you may end up with a "dud." The questions are how it performs and how it looks and how many miles you can get out of it. Of course, technology impelled the sports bras forward: petroleum-based fibers, for example spandex, were crucial to developing comfortable models. The first Jogbra was cotton with some spandex, and it was an improvement over nothing, but it stayed wet and therefore was more abrasive. Lawson commented on today's sports bras: "How privileged we are to enjoy exercise with these new materials."[34]

Treadmills, Cameras, Sensors, and the Biomechanics of the Breast

At first, sports bra designers built on personal experience and anecdotal evidence to create bras that met the demands of increasingly active athletes. By the late 1990s, testing to produce better bras—by which they meant sports bras that were more comfortable, supportive yet lightweight, used wicking fabrics, and avoided chafing—had moved to the laboratory. Scientists used treadmills, a series of computer-linked sensors, and cameras to evaluate how the breast moved during exercise and which designs best supported the breast tissue. Their samples grew from a handful of subjects to several dozen and involved the study of the naked breast as well as of athletes wearing bras of various kinds.

Kinesiology developed in the 1960s in English-speaking countries, initially an Anglo-Australian-Canadian-American science that has spread to Europe. The field has grown to at least 112 recognized journals[35] with articles that address topics from physiology to mechanics, from training to community-based programs, from accidents to treatments. Some of the journals are decades old. The first, the *Journal of Sport and Exercise*, inaugurated in 1979, published an article that examined and confirmed bias against women concerning their motor tasks.[36] Kinesiology departments enable students to earn Bachelor's and Master of Science as well as doctoral degrees, with the first founded at the University of Waterloo in Canada in 1967.

Regarding the sports bra, specialists studied the kinesiology of the breast. Female breasts are primarily composed of glandular tissue and fat, held in position by the delicate Cooper's Ligament. Any excessive amount of breast movement or bounce, as happens during exercise, puts strain on these ligaments and can cause them to stretch. In the long term this leads to breast sag.[37] Because the female breast does not contain strong intrinsic structural support, breast motion is difficult to reduce. It has been suggested that the skin covering the breast may also act as a support structure for the breast, but this has not been quantified.[38]

Inadequate breast support, coupled with excessive breast movement, is the most likely cause of sore and tender breasts after exercise. For example, the average size 36C breast is estimated to weigh about 10 ounces. Researchers calculate that a breast of this size, with minimal support, running at about 5.6 mph, will bounce as many as 4.7 inches up and down, relative to trunk movement. A good sports bra can decrease this by about one-half or more.[39]

But pain is not linked to the size of a woman's breasts. Because of discomfort an A-cup woman could be prevented from doing sport just as much as a woman with double FF cup size. Breasts can hurt for three main reasons: tenderness during the menstrual cycle; permanent breast pain; and exercise-related pain due to stretching of the breast tissue. Only a sports bra, not a regular bra, can reduce motion and provide support for exercise, as several decades of study have demonstrated.[40]

Early designs in the burgeoning market sought to avoid conventional bra construction through innovations based on scientific investigation. Lawson, who later headed Champion's sports bra division, noted that some early designs turned the cups upside-down in search of stability of the breast, used inappropriate fabric, and were often produced by athletic supporter companies that quickly left the bra business. These bras, like the Lady Duke, featured thick, heavy-duty fabrics that were "not pretty." As noted, some lingerie manufacturers also joined the business with insufficient scientific understanding and quickly left it after producing inadequate sports bras.

According to some sources, the early designers turned to Dr. Christine Haycock for advice.[41] It has been hard to determine precisely Dr. Haycock's contribution to this story, but after a distinguished career in the US Army during World War II in the US Cadet Nurse Corps, she became the first woman intern at Walter Reed Army Medical Center, and after a distinguished career retired in 1984 at the rank of colonel. As a diplomate of the American Board of Surgery and a fellow of the American College of Surgeons, Dr. Haycock was a leading expert in Sports Medicine and was a member of the American College of Sports Medicine. She published widely in this field, particularly concerning women in sport.[42] In the 1970s and 1980s she conducted pioneering research that measured breast movement by observing women running on treadmills. Coaches and other sports officials were most concerned about women's potential for injury. Haycock was concerned with solving the problem of soreness because, as a softball player, she experienced breast pain every time she pitched a game. An associate professor of surgery at the University of Medicine and Dentistry of New Jersey, she represented the needs of female athletes by serving on the American Medical Association Committee on Medical Aspects of Sports. She also conducted key studies investigating breast support and protection.[43] She said, "On the one hand, the studies were debunking the myth that women shouldn't be in contact sports because they'd hurt their breasts. But on the other hand, the research gave credence to women's complaints of breast pain." In one study she measured the force of a breast in motion. When bra manufacturers started looking at this research and asking Haycock for advice, she suggested wide bottom bands for extra support and straps that were not so elastic that they let breasts bounce. Several "intimate apparel" companies responded by making "beefed-up everyday bras built for sport—not exactly ideal gear, but a sign that the needs of female athletes were moving into the mainstream."[44]

Dr. Lawson first became interested in studying female physiology and the sports bra in 1977 after she completed her first marathon with shoulder pain and chafing. Her master's thesis at Utah State University in 1984 quantified the comfort and support problems for women athletes in the absence of good sports bras. She used a sample of sixty women, fifteen each in cup sizes A to D, to record motion of the breast and measure perceived comfort in treadmill running both of nude subjects and those wearing bras. Unlike researchers today who have access to computers with USB ports, inexpensive video camera plug-ins, and immediate results recording, Lawson had to use 16mm cameras with black and white film that had been rented from Hollywood studios. She recalled how difficult it was to get volunteers to participate because of conservative attitudes in Utah among women and the fact that her subjects had to "wear" a one-inch sticker with a black crosshair on the nipple; issues of modesty prevailed.[45] The study attracted the interest of Jogbra, whose representatives contacted her and which led to one of her first publications in 1987.[46]

Lawson continued her academic research at Oregon State University in the laboratory of biomechanics, where technical advances made studies easier to conduct and more complete. Those changes included partial digitization of tests, the use of six to eight cameras, three-dimensional results, real-time recording, and infrared analysis. Over time Lawson and her colleagues observed subjects with DD and DDD-sized breasts, although those individuals were hard to recruit. They learned to extrapolate for changes in breast mass over 32-, 34-, 36-, and 38-inch busts and different cup sizes to design appropriate sports bras. The most important early result was simple and profound: vertical displacement work was far more important because 90 percent of the ground reaction force was vertical and only 10 percent was horizontal.[47] Researchers like Joanna Scurr in Portsmouth, England, as I discuss later, are currently studying horizontal surge and how it adds to discomfort.

In a 1990 study, Lawson and a colleague compared seven sports bras on quantitative and subjective measures of support and comfort for small-, medium-, and large-breasted women during exercise. Sixty subjects representing A, B, C, and D cup sizes were filmed while jogging on a treadmill at 6 mph in each bra style and in the nude condition. The average vertical displacement of the breast relative to the body was calculated for each condition. Subsequently, subjects completed questionnaires after engaging in vigorous exercise in each bra to assess subjectively comfort and support for each of the seven sports bras. Lawson noted significant differences by bra style in subjective mean comfort scores and a significant interaction between subjects' cup sizes and subjective mean support scores. Lawson tracked the motion of the breast to the smallest detail; breasts tended to move in a figure-eight pattern that reflects the arm swing and shoulder rotation during running. Lawson said, "We can tell you how fast the nipple is going as it changes direction. We can calculate velocity and acceleration." She added, "When you have a breast traveling downward at a pretty good rate of speed, and then you change direction as your body hits the ground and begins to rise up again, you can get some pretty drastic acceleration as the breast speeds up to try to catch up with the rest of the body. The point of the sports bra is just to slow all that down."[48] Generally, bras that effectively controlled breast displacement tended to score lower on comfort. There were, however, exceptions to this trend, and no single bra at this stage of development rated well on all measures of performance.[49]

Improvements in sports bra design based on science were relatively slow in coming, but gained momentum as more women and girls joined athletic activities and demand for better sports bras grew. This may be a distinction with other "feminine" technologies whose development was male-dominated and thus presented barriers to safer devices and appropriate technologies—such as a birth control pill that was safer and more efficacious that took years to produce.[50] Women were not included in many medical and other clinical trials on a regular basis until the 1990s.[51]

According to a 1999 Australian study, breast pain was common during exercise in up to 56 percent of subjects. In an attempt to analyze the movement of the breast and the resulting pain, the movement of the female breast tissue was quantified in four conditions of breast support (sports bra, fashion bra, crop top, and bare-breasted) during four different activities (running, jogging, aerobics march, and walking), although in research involving few subjects. The results showed that wearing external support for the breast tissue reduced absolute vertical movement and maximum downward deceleration force on the breast. Support also reduced perceived pain: the sports bra provided superior support for the breast in relation to the amplitude of movement, deceleration forces on the breast, and perceived pain.[52]

Deirdre McGhee, Julie Steele, and their colleagues at the Biomechanics Research Laboratory at the University of Wollongong in New South Wales, Australia, have spent several years studying how best to ensure that women who engage in exercise wear the proper sports bras. In one study McGhee and her colleagues aimed to determine the bra-breast forces generated in women with large breasts while these women wore different bras with different levels of breast support during both upright standing and treadmill running. The forces exerted on the breast during exercise are rather large. The mean bilateral vertical component of the bra-breast force in standing was 11.7 ± 4.6 N, whereas during treadmill running the mean unilateral bra-breast force ranged from 8.7 ± 6.4 N to 14.7 ± 10.3 N in the high and low support conditions, respectively (1 Newton is the amount of net force required to accelerate a mass of one kilogram at a rate of one meter per second squared). The researchers also determined that, logically, breast mass was significantly correlated with vertical breast displacement in the high support condition. Thus, manufacturers must take into account the wide range of breast masses of women with large breasts to ensure their sports bras can reduce force generation and breast discomfort by providing a high level of breast support while these women participate in physical activity.[53]

The Wollongong biomechanics research team has studied how best to support breasts during exercise. They affix light-emitting diodes to women's clothing and chart them running on treadmills; the women run with and without bras. Computer systems track the breasts' motions in three dimensions and the researchers can see how and where they move and how movement is affected by bras. Breasts move in a sinusoidal pattern, they move a lot, and if the bra is not appropriate the breasts move large distances and leave the cups. The larger the breasts and the more they move, the more momentum they generate. Controlling this momentum requires a large force, usually applied through bra straps. When straps are thin, the pressure exerted through them can be so great as to leave furrows in the shoulders of large-breasted women. As the straps dig into the brachial plexus, the nerve group that runs down the arm, they may cause numbness in the little finger. In some cases, breasts can slap against the chest with enough force to break the clavicle.[54] The long-term consequences of this are only now being studied. To limit motion, compression bras are more popular but less comfortable: They squash the breasts against the body, thereby reducing the amount of weight the bra has to cantilever.[55]

Another major research venue is the University of Portsmouth, England, where Dr. Joanna Scurr works on improving bra design. Scurr is a Reader in biomechanics and the biomechanics division leader. Her research interests are in the area of breast health. Scurr established a Research Group in Breast Health in 2008. The group consists of five members, including research assistants and PhD students, whose focus is applied research, fundamental research, and clinical support. Their research began in 2005 with a major research project on the biomechanics of the breast and the effectiveness of sports bras at reducing breast motion. Scurr's team noted that breasts bounce more during exercise—up to 21cm rather than the maximum 16cm bounce measured in past studies. Bras are designed to stop breasts bouncing up and down, but Scurr's work showed that breasts also moved side to side and in and out. Significantly, Scurr's latest studies have found that breasts move as much during slow jogging as they do at maximum sprint speed. She said, "This makes wearing a sports bra as important if you jog slowly as if you sprint."[56]

In an article published in *Applied Biomechanics*, Scurr and colleagues presented the results of research comparing breast support requirements during over-ground and treadmill running. They investigated three-dimensional breast displacement and breast comfort during exercise. They fitted six female D-cup participants with retro-reflective markers placed on the nipples, anterior superior iliac spines, and clavicles. Five ProReflex infrared cameras (100 Hz) measured three-dimensional marker displacement in four breast support conditions for various running trials and speeds. They gathered subjective feedback on breast discomfort using a visual analog scale. Their findings indicate that breast motion studies that examine treadmill running are fully applicable to over-ground running.[57] In another study, they noted that bare-breasted kinematics, not surprisingly, significantly increased with cup size during running. Differences in breast displacement, velocity, and acceleration between cup sizes could be predicted using estimates of breast mass based on conventional brassiere sizing. These data inform the design and evaluation of effective bra support.[58] Because breasts move far more than ordinary bras are designed to cope with, like her other colleagues in breast mechanics, Scurr is working with major bra manufacturers in Britain and globally to design bras that lessen movement in all three dimensions and reduce much of the pain many women suffer when exercising. Militaries have also become customers of research in their effort to provide the proper attire for female soldiers.[59]

The research of Scurr and colleagues also questioned the finding that compression bras are better for reducing movement in small-breasted women and encapsulation bras better for larger-breasted women. She found instead that encapsulation bras are better at reducing breast movement in women of all cup sizes. She said the big question is why we know so little about the movement of breasts. "Sports science has always been dominated by men and for them, studying breasts is seen as slightly laughable. For women, though, it's completely credible—they can see the benefits." She continued, "At conferences when I am asked what I study I say 'bouncing breasts' rather than breast biomechanics. It makes people laugh nervously but they always want to know more. So little has been known about this subject until recently."[60]

McGhee and Steele conducted a study to determine whether a sports bra designed to both elevate and compress the breasts could decrease exercise-induced breast discomfort and bra fit discomfort experienced by women with large breasts relative to a standard encapsulation sports bra. They compared breast kinematic data, bra fit comfort, exercise-induced breast discomfort, and bra rankings in terms of preference to wear during running among twenty women with large breasts who ran on a treadmill under three bra conditions: an experimental bra that incorporated both breast compression and elevation, an encapsulation sports bra, and a placebo bra. They concluded that the experimental bra worked best through greater breast elevation and compression, and hence was better than a standard encapsulation sports bra during physical activity for women with large breasts.[61]

Yet even with extensive study, many women have purchased sports bras based on insufficient information to select an appropriate one, perhaps because of contradictory findings. The fact remains that many women do not buy proper fitting bras with adequate support. This can lead to musculoskeletal pain and inhibit women who participate in physical activity. McGhee and Steele undertook a study to determine systematically the best method for women to choose a well-fitted bra on their own. The bra size of 104 women (mean age = 43.5 ± 13.2 years; average bra size = 12DD (34DD); band size range = 10–24; and cup size range = A–G cup in Australian sizes, which are generally close to British and US systems) was determined through self-selection and bra size measurements. The researchers then compared this to the "correct bra size" as determined by professional bra-fitting criteria. Fully 85 percent of the participants chose ill-fitting bras and bra sizes as determined by self-selection or using the bra-sizing measurement system. They called for education of professional bra-fitting criteria for medical practitioners and allied health professionals to assist their clients or patients in choosing a well-fitted bra.[62]

McGhee and her colleagues suggest that adolescent athletes should receive better information, perhaps a well-written pamphlet, to ensure they select their sports bras properly from an early age. They worked with 115 adolescent females from four regional sporting academies aged 16 years old and with an average Australian bra size of 12B (34B), the experimental group of which received an education booklet on bra fit and breast support from a sports physiotherapist, while the control group received no intervention. They used a questionnaire to evaluate results. Four months after receiving the education booklet, the experimental group had improved their bra knowledge significantly more than the control group and had a good sense of bra fit and breast support. They concluded that "bra knowledge, bra fit, and level of breast support in adolescent female athletes were all poor but improved significantly after receiving an education booklet about breast support designed specifically for them."[63]

The major actors in development of the sports bra have researched both the physiology (sweat patterns) of female athletes and the biomechanics as well as what type of support the female athlete needs based on the impact level of her sport or activity. Studies of breast biomechanics at the Nike Sport Research Lab confirm that breasts move in a figure eight pattern that varies by activity and level of intensity. Researchers have also studied how low-impact activities like yoga, walking, and golf apply 1 to 1.5 times body weight in impact forces to the body; aerobics, dance, tennis, and running, 1.5 to 3 times body weight; and agility drills, jumping, and landing apply forces anywhere from 4 to 10 times the individual's body weight. After study of the key sweat zones of female athletes, designers determined that a crucial consideration is to manage sweat away from the chest and back, where women sweat the most in terms of their sports bras. The major companies have spoken with thousands of female athletes—from high school and collegiate players to elite athletes including members of the US national teams to understand exactly what these athletes look for in a sports bra and how to help them find the right bra for their needs.

Another innovation was the introduction of a flexible underwire. Until the mid-1980s, manufacturers used underwire primarily for large-breasted women. The wire, sewn into the seam beneath each bra cup, gives more structure to the cup and sides of the garment. But most bra manufacturers used "a merciless U-shaped cord of stiff steel." If the wire didn't fit the U perfectly, or if machine-washing distorted the U, then the underwire might hurt the breast or dig into the rib cage. In the late 1970s, S&S Industries introduced a lighter, plastic-coated wire that provided flexible support and served women whose breasts were of all sizes. For women with smaller breasts, the underwire allowed design of a bra without padding, while underwire continued to provide support for larger-breasted women.[64] Nike created a patented molded, flexible under-bust support that actually reduces the overall weight of the bra without compromising support and stability.

Ultimately, all sports bra designers learned to focus on providing vertical, horizontal, and surround support in four key areas for all the bras in the collection: the chest band, which is the foundation for good support and anchors the bra to the body; the cups that surround and support the breast separately and/or together; the shoulder straps, to ensure lifting support for the garment in the front and stretchable fit in the back for full range of motion; and the wings, for supporting the side of the cups and providing a flexible fit for different sport activities.

Increasing Breast Size

The early sports bra model, largely the pullover shelf-bra design "established during the aerobics craze worked reasonably well for A and B cups but was woefully inadequate for D-plus sizes." Renelle Braaten, a Montana hairdresser who enjoyed racquetball and volleyball, found great difficulty playing because her DD-sized breasts "bounced all over the place." In 1988 she and her mother, an accomplished seamstress, worked together to design a bra for large-breasted women that was strong, used a relatively non-elastic material, employed a crisscross design for better back support, and wide straps. She used both front and back closures, at first with zippers, but settled on hooks to achieve a very snug fit. Braaten attempted to sell her design to other companies, but they all refused because they could not see a demand for it. They doubted that many large-breasted women would engage in active sports. She therefore determined to create her own company, Enell Incorporated, founded in 1993, that grew rapidly and whose corporate offices frequently get letters and emails from grateful women. (Braaten talked Oprah Winfrey into trying the Enell bra on TV which stimulated such sales that an eight-month inventory was gone in three days.) To this day, Braaten makes a special effort to reach adolescents and teens with a large breast size so that they feel they can continue in sport and not be embarrassed by their breasts; the "Enell" enables participation with support so that other proper sports mechanics prevail.[65]

Enell's blogspot enables "big girls" who have never been called "svelte" to share their experiences on the road in comfort. They discuss the challenges of finding a sports bra that is simple to put on, not one that requires the assistance of a "crowbar." Enell claims that one of the major improvements in their designs concerns the straps. For many women bras with adjustable straps create problems—they often don't stay where women want them or "end up stretching past a point of being able to support the breast adequately. This leads to either the dreaded droop or the incredibly frustrating shoulder strap slip." But the design of the Enell straps (and of straps of other such bras made for large-breasted women) eliminates the possibility of slipping, ensures that the breast remains properly lifted and supported, provides comfort during high-impact activities, and evenly distributes the weight of the breasts over a wider portion of the shoulder.[66]

The larger the breasts, the more they move and the greater the discomfort. A runner wrote me that a friend of hers "back in college [in the late 1970s] was a 36E and she had a system. . . . She would buy a smaller size bra in a cotton fabric that didn't stretch at all, then she would put two t-shirts on with the bra over those. Breathing was a challenge, and she eventually got a breast reduction!"[67] But the design improvements in the 1990s made it possible for big-breasted women to find a sports bra with adequate support and little bounce. According to one source, more and more women need bras that offer greater support than those engineered for B cups; she writes that poor eating habits, breast implants, and the estrogens in birth control pills have led to an increase in the past fifteen years of more than one bra size for the average American woman.[68] Americans in general are gaining weight; ten years ago 34B was the average breast size; now it is 38C. This means that larger-breasted athletes are no longer a niche market. Big corporate players are catering to C and D sizes and producing sport-specific bras, from styles for low-impact activities like yoga and walking to styles for running, and with bras that combine compression designs with encapsulation to support and flatter.[69]

Jen, a 40-year-old mother of two, recently published a *Huffington Post* lament about the failure of bra manufacturers to meet her workout regimen needs—and at the same time to make an attractive bra for a woman who is a 38DDD. Jen claims there are no sports bras out there for large-breasted women. She likened one to a "t-shirt with a bit of elastic around the edges," claimed that another style "caters to the boobless," and criticized virtually all of them for being difficult to get on and off.[70] One would think that common sense would lead to creation of a sports bra for women like Jen that had back clasp. But consider how far the sports bra has evolved since its creation in the 1980s.

Shock Absorber, a UK company, established the Shock Absorber Sports Institute (SASI) in response to what it claimed was a gap in research on large-breasted women. The company noted that "previous breast movement research . . . only monitored smaller cup sizes, only looked at the effects of vertical breast movement, only analyzed one stride during running, and then only up to 7.5 mph." SASI focused on larger breasts, higher speeds, and longer time-motion studies to examine breast motion vertically, from side to side, and forward and backward using a series of cameras through a series of activities (static, jumping jacks, and on a ramped treadmill) to determine comfort levels. This enabled designers to reduce bounce by up to 74 percent, that is, in minimizing breast movement compared to a "normal" bra.[71] According to SASI 73 percent of women who exercise regularly do not wear sports bras. Researchers determined that among size 34A breasts, breast movement ranged up to an average of 1.5 inches away from the resting place of the body. Shock Absorber has an online "flash-based boobies-physics simulator"; plug in a cup size and a level of activity, and it produces a 3D animation of breasts of that size bouncing free, bouncing in a regular bra, and hardly moving at all in one of their sports bras.[72]

In an earlier study the Wollongong group determined that exercise-induced vertical breast displacement and discomfort in women with large breasts were reduced during deep water running compared to treadmill running. They worked with sixteen women (mean age = 32 years, range 19–43 years; mean mass = 74.1/kg, range 61–114/kg; mean height = 1.7/m, range 1.61–1.74/m), who were professionally sized to wear a C+ bra cup. The study group ran at a self-selected stride rate on a treadmill and in 2.4/m deep water. Immediately after running, the subjects rated their breast discomfort and breast pain (visual analog scale) and their perceived exertion (Borg scale). The exercise-induced breast discomfort was significantly less, although the perceived exertion was significantly greater during deep water running relative to treadmill running. Although there was no significant between-condition difference in vertical breast displacement, mean peak vertical breast velocity was significantly less during deep water compared to treadmill running.[73]

A 58-year-old self-proclaimed "top-heavy runner" told me she can "vividly remember how disappointed [she] was when [she] tried on [her] first Jogbra." She writes, "In fact, [in] my first two marathons, the Tampa/St. Pete in 1985 and NYC in 1987, I wore a Jogbra *plus* I wrapped an ace bandage around my chest so I would not bounce." She has not found a sports bra that is adequate from the major well-known companies, but found "a small woman's athletic sporting company . . . that designed and sold a sports bra that I wore for many years, called a Frog Bra. [It is a] great company run by women for women. And more recently, you have companies like Lululemon Athletica, a Canadian company that has added a great deal of style and upgraded fabrics to athletic wear and sports bras." She adds that "equally important to my running shoes is my running bra. Maybe even more important."[74]

Materials Science

Specialists publish extensively in such journals as *Fibers and Polymers* and *Textile Research Journal* on a variety of new materials and fabrics that found their way into sports attire applications. These included polyester, spandex (anagram of "expands"), Lycra (introduced by the DuPont Company in 1959), and CoolMax in various mixes, often with some cotton. These materials had a revolutionary impact on attire generally and sports bras in particular in the way they improve comfort and support, wick away moisture (sweat), and minimize friction. Publications on these materials date to the 1980s but specialists from industry and from trade journals began to publish much more extensively in the 1990s and beyond, and a variety of manufacturers have expanded factory capacity to produce the fabrics in large quantities to meet growing medical, scientific, sports, and other demands. Manufacturers use ever-more-smooth microfibers to reduce friction. In some cases the straps are specially engineered, some with gel inside for added cushioning.

Materials science, nutrition, and sports physiology have joined together to improve athletic performance generally in the postwar world. Bob Richards, the second person to pole vault 15 feet (4.6 meters), won two Olympic gold medals in 1952 and 1956 and claimed that few of today's vaulters could achieve his heights with a bamboo pole. Instead, with fiberglass composites they have gone over 6.0 meters. In 1964, Jim Ryun, a high school student, ran the mile in 3:59 on a cinder track; today's sprinters, hurdlers, middle- and long-distance runners have all-weather tracks engineered for less muscle fatigue, consistent bounce, and significantly faster times, with Hicham El Guerrouj having run a mile under 3:44. Tennis rackets made of composite materials are lighter, bigger, and with a larger "sweet spot" than those made of wood. Skis with deep side cuts and hourglass shapes enable athletes and novices to go faster with greater maneuverability—but with greater risk than skiers only forty years ago.

The evolution of materials used in the industry is not linear. Some of the earliest fibers used—like cotton and nylon—are still used today in some cases based on materials price and preference of the manufacturer. Nike moved from using blends of cotton/Dri-FIT/spandex in the first compression bras to incorporating zones of strategically placed mesh and the use of Dri-FIT Denier Differential material (a unique two-layer fabric structure with large and small yarns that quickly moves sweat away from the skin to the surface of the garment so you feel drier) to blends of cotton/Dri-FIT/spandex beginning in 2011. Making a sports bra is complex, with careful consideration of multiple layers of materials and how those materials will work together to provide comfort, moisture management, and support.[75] That is, construction must incorporate stretch and nonstretch zones in a responsive, breathable fabric that pulls sweat away from the skin for dryness and comfort. Materials should reduce the potential for chafing in areas where the athlete is prone to chafe due to the garment rubbing against her skin—under the chest and underarm. For example, spandex is used primarily for stretch and comfort, in both the encapsulation and compression of the breasts as it blends with other materials to provide the right amount of stretch with support, and also to aid in range of motion throughout the garment's fabric.[76]

Along with advances in support, sports bras now offer high-performance moisture management; and technological advances made for outdoor sporting goods like parkas and tents have crossed over to sports bras. Welded and molded parts have replaced stitched seams, which were too often the source of chafing and discomfort, although as noted earlier the welded parts have not always "seamlessly" entered the bras! The first high-support seamless bra, introduced in 2003 by Champion (Jogbra's current owner), combined slick nylon fabric with seamless design for friction-free performance. Champion's Vapor bra uses Cocona fabric technology, a superwicking fiber made from coconut shells that outperforms CoolMax.[77] Like many other fibers coming now from natural materials rather than petroleum products, the coconut fiber is strong and durable and is mixed with activated carbon and recycled polyester to form an odor-absorbing and moisture-wicking fabric.[78] One bra may include eight types of fabrics to vary the amount of support, sweat-wicking properties, and ventilation levels.

Like most companies, New Balance shod athletes before dressing them. They entered the market to sell shoes, not apparel. But although in the early years there were just a few players, today there are dozens of sports bra companies—and companies that sell a range of products including sports bras. Not only are there many players, but, a New Balance spokesman said, "Women have been more educated on protecting breast tissue and ligaments. Activities, aging, and children all have an impact, so the bra is very important." Structure has evolved and fabrics have evolved. Cotton remains an important component, but it has moved forward in combinations and amounts of poly and Lycra for more or less support, more or less rigidity, more or less moisture wicking. Better knowledge has led to more scientific use of elastic, orientation and kind of seams, and nature of threads. Body type dictates style in many cases—whether pullover bras, zip in front, clasp in back, size and nature of straps.[79]

For some women athletes the materials make all the difference. A sub-3 hour runner says that "the biggest criteria for me is having one that fits well but is not too tight and doesn't constrict the boobs, rib cage or arm motion. The worst feeling is having a bra that comes up too high on the sides and chafes below the armpit. I hate that. I prefer bras with dryfit material." This runner has been competing since she was 13 years old, so has been wearing sports bras since the mid-1990s. She writes, "The first few were cotton and eventually got worn down and got holes . . . or the elastic would soften. Materials are definitely a lot better now. I don't like the ones with built-in cups or bras. Being small chested, I don't really have to worry about not getting enough support."[80] Another marathoner I know writes she is "one of those small breasted women who doesn't really need to be picky about sports bras. I literally had four identical ones that I bought at K-Mart for under $10 each that I ran in for about a decade. Fancier ones chafed me pretty badly." She remarked on the improvement on the technology, "In a recent triathlon, I decided to wear just a sports bra for the bike and run, which meant that I had to pony up and get something better looking. The technology had clearly changed over the last decade, because I got four new ones, and they don't chafe at all, unless I wear the same one two days in a row, particularly after a long run."[81]

Lawson pointed out how women's attitudes toward sports bras have changed in the forty years she herself has been running and biking. Worries about a slipping bra strap have passed. Women have lost their self-consciousness and can choose models according to their desires. Designers can focus on other issues. For example, "modesty" has recently become a buzzword. Increased blood flow, cooling, and other factors make erect nipples show. While some women do not care, others are concerned, and manufacturers have added "modesty" padding. But the "Holy Grail" of the sports bra was one that masked the nipple yet did not make it too hot, while adding more fabric meant problems of airflow and moisture. Younger women also seem to want more shaping. All of this has added to interest in new materials, fibers, and fabrics. Miller's sense is that product engineers' understanding of physiology lagged somewhat behind the development of new materials and mixes of materials.[82]

Gender and Market Demand

Hinda Miller recalled, "In the early '70s, Title IX of the Equal Education Amendment mandated that institutions receiving federal funding had to spend equal amounts of money on men's and women's programs. By the late '70s, over 6 million women were running, hooked on the feeling of health, strength, and endorphins. I was a runner, along with my soon-to-be partner Lisa Lindahl. . . . Sportsbras now generates $500 million at retail and is recognized as having as big a role as Title IX in increasing women's participation in sports and fitness."[83]

According to the Department of Education, Title IX covers state and local agencies that receive federal education funds. These agencies include approximately 16,000 local school districts, 3,200 colleges and universities, and 5,000 for-profit schools as well as libraries and museums. Programs and activities that receive these funds must operate in a nondiscriminatory manner. These programs and activities may include, but are not limited to: admissions, recruitment, financial aid, academic programs, student treatment and services, counseling and guidance, discipline, classroom assignment, grading, vocational education, recreation, physical education, athletics, housing, and employment.[84] Title IX governs the overall equity of treatment and opportunity in athletics while giving schools the flexibility to choose sports based on student body interest, geographic influence, budget restraints, and gender ratio. In other words, it is not a matter of women being able to participate in wrestling or that exactly the same amount of money is spent per women's and men's basketball player.[85]

In 2010, the administration of Barack Obama reversed a 2005 policy instituted under President George W. Bush that allowed schools to use a survey to prove a lack of interest in starting a new women's sport; the Bush administration encouraged schools to consider a nonresponse to the questionnaire as disinterest. This was an attempt to abandon the enforcement capability of the federal government under Title IX to ensure equal access and to allow the regulated universities to regulate themselves.[86]

According to the NCAA, before Title IX, fewer than 300,000 high school girls—one in twenty-seven—played sports and fewer than 32,000 female athletes participated at the collegiate level. By 1974, just two years after the passage of Title IX, the number of high schoolers participating in sports had skyrocketed to 1.3 million. In 2011 more than 3 million high school girls—one in two—played sports, and in 2010–11 more than 191,000 females played a wide variety of NCAA sports. All of this has led to ever-growing demand for the sports bra. (And, on another note, female athletes had significantly higher graduation rates compared with the overall student population.)[87]

A gender gap, wage gap, and technology gap persist in the twenty-first century, the achievements of women in society as country leaders, as professionals in literature, culture, and sport notwithstanding. After some noteworthy successes in the US courts in removing long-standing obstacles to equality, for example, the courts—and many political leaders—have backtracked. In 1965 in *Griswold v. Connecticut*, the US Supreme Court confirmed a woman's right to medical information, including birth control, as essential to privacy. In 1973, *Roe v. Wade* again confirmed a woman's right to privacy in medical decisions with her doctor, in particular those concerning birth control. In *Automobile Workers v. Johnson Controls, Inc.* (1990), the Court blocked the effort of Johnson Controls to bar female employees from certain jobs. In this case, the company adopted a patriarchal and sexually discriminatory violation of Title VII of the 1964 Civil Rights Act when it forbade women of reproductive age from working in a battery foundry where they might be exposed to high levels of lead—which the company assumed would damage any fetus. But "Johnson's fetal-protection plan fell outside the bona fide occupational qualification exception of Title VII, since the exception only permits employers to discriminate based on qualities that detrimentally impact on an employee's job performance."[88]

The courts and state governments have again begun an assault on women's rights to technological equality. In the 2010s a dozen states have passed laws to give fetuses more rights than women by closing access to medical clinics that perform abortions, but in fact these clinics generally provide other, mostly crucial public health services, primarily for poor women and children. And the

Supreme Court affirmed the right of employers to discriminate against women. Over her nineteen-year career at Goodyear Tire, Lilly Ledbetter was consistently given low rankings in annual performance-and-salary reviews and low raises relative to other employees. She learned that she was making significantly less money than any male working in her department: $3,727 per month, compared with $4,286, the salary of the lowest paid male employee doing the same job (a difference of $6,708 per year). She sued Goodyear for gender discrimination in violation of Title VII of the Civil Rights Act of 1964, alleging that the company had given her a low salary because of her gender. A jury found for Ledbetter and awarded her more than $3.5 million, which the district judge later reduced to $360,000. Goodyear appealed, citing a Title VII provision that requires discrimination complaints to be made within 180 days of the employer's discriminatory conduct. Ignoring the fact that she had been discriminated against, and ignoring "current effects," the Court found a way to find for Goodyear. By a 5-4 vote the Court ruled that Ledbetter's claim was time-barred by Title VII's limitations period. The opinion by Justice Samuel Alito held that "current effects alone cannot breathe life into prior, uncharged discrimination."[89] Will states pass laws to prevent women from having access to sports bras as somehow not in keeping with their status as breeders and mothers?

Having access to performance-oriented sports bras may not have made women more athletic, but it definitely enabled them to achieve greater personal results with greater personal comfort and perhaps therefore helped them become more independent and less subject to discrimination. Lisa Lindahl observed, "The history of the sports bra really tracks with the history of women in athletics." Prior to Title IX, women played half-court basketball, and running more than a mile was seen as a big accomplishment. As women upped the intensity of their sports from the mile to the marathon, they demanded equipment equal to the task. In turn that equipment made it possible for women to go longer, harder, and faster. "It's been a transformative piece of clothing," observed LaJean Lawson. "Finally, we have what we need to play hard, play well, and be comfy."[90] The sports bra served the forces of social change.

The Title Nine Company produces athletic clothing for female athletes, including sports bras that are sized by cup and rated according to barbells, with five barbells providing maximum support during high-impact activities.[91] A tour of the US Patent and Trademark Office reveals dozens of patents in the 2000s for sports bras with adjustable support systems, such as CoolMax support structure for large-breasted women. This indicates that, in response to a growing market and increasing numbers of women getting involved in high-impact sports, modifications toward comfort and support of the sports bra will continue to meet the needs of female athletes.[92]

The materials should be comfortable, have wicking capability, and be "breathable" even during long periods of sweaty exercise. Adjustable, shaped, padded straps are a plus; some straps contain gel for comfort. Molded cups provide shape and modesty. Side panels help provide support. A larger strap underneath the breasts provides stability; this is a solution for those women whose bras have caused chafing or permitted too much bounce.[93] There are scores of websites as well as forums in the major runner's magazines to assist women in choosing the right sports bra, so it is clear that women athletes need to approach the purchase of a sports bra carefully—there's no avoiding going to a shop or a marathon expo to try on several styles and manufacturers' models to ensure comfort, support, and fit. At running expos around the world it is possible to find a saleswoman who really knows what she is talking about—and who understands the biomechanics of the breast.

In this way, researchers in Europe, the United States, and Australia have changed bra design into a science based on empirical research that employs larger samples of women in longer term studies, with important implications for women of a wide range of breast sizes and activities. The research has been engaged by the military, national health programs, and lingerie manufacturers. Given these changes and these numbers, as Hinda Miller pointed out, it was natural for her to help create an iconic piece of women's underwear that has found its way into museums: the Smithsonian Institution and Metropolitan Museum of Art have Jogbras somewhere in their collections. Miller says that the sports bra equally is an icon of the feminist movement in the boardroom and playing field—as her own career as a businesswoman indicates.[94]

In the sports bra, as we have seen, many issues of gender came together. Women athletes recognized that a male-dominated sports science had failed to give them their due, nor had athletic apparel companies considered producing attire appropriate for them, nor had the male-dominated national and international Olympic committees considered that women were capable of such sports as marathon running. In the 1980s sports physiologists, many of them women, turned to the study of breast mechanics and physiology, the better to design sportswear for increasingly active women. And progressive federal laws and regulations intended to ensure equal access at least to federally sponsored programs led to rapid increases in the number of women in colleges and universities and in athletic programs.

Gender, science, technology, and consumer demand came together to enable the development of the sports bra—a combined achievement of two young female graduate school inventors, of the women's movement, of material scientists working with new fabrics that support the muscle and tissue during exercise and wick moisture away from the skin to prevent chafing and bruising, of growing numbers of female athletes who performed at high levels, and of federal programs to end systemic discrimination against schoolchildren and university students on the basis of gender.

Researchers in the banana ripening room, 1956. The banana, a staple for many people, ripens slowly and ships well, so that it has become a commonplace fruit even in North America and Europe, far from the tropical plantations where it is grown. United Fruit Company Photograph Collection, Baker Library, Harvard Business School (olvwork729996).

3

Sugar, Bananas, and Aluminum Cans

Technology, Colonialism, and Postcolonialism

I have ransacked the encyclopedias,
And slid my fingers among topics and titles,
Looking for you.
And the answer comes slow.
There seems to be no answer.
I shall ask the next banana peddler the who and why of it.
Or—the iceman with his iron tongs gripping a clear cube
in summer sunlight—maybe he will know.
—CARL SANDBURG, "OLD-FASHIONED REQUITED LOVE"

I love bananas. They provoke memories, not of unrequited love, but of childhood, fruit bread, and running marathons. Bananas are easy to digest, high in
potassium, and athletes love them. At the St. John, Canada, Marathon a few
years ago, so many marathoners descended on the city that there was not one
banana to be found almost days before the race. In a way, this is not surprising,
since bananas are not indigenous to the Canadian maritime provinces. Bananas
come from Costa Rica, Honduras, Ghana, and Jamaica, and it seems a miracle
that I recently ate one in Tomsk, Siberia, near the world's largest swamp. I also
enjoy Carl Sandburg, his biography of Abraham Lincoln, and his remem-
brance of the banana peddler.

But bananas are not just bananas—and sources of memories and pleasures.
Bananas are technologies. They are plantations of trees, irrigation systems,
workers, trucks, trains, and ships, urban infrastructure, and hospitals. They
are tools of multinational corporations that use political, economic, and other
cudgels to enforce unequal trade relations. They are roads and pulleys and
conveyors. They are stems—hybridized and selected to facilitate growing, harvest, local handling, packing, and long-distance shipping. The United Fruit
Company put tremendous pressure on the Jamaican independent growers and

those in other countries to produce bananas free of blemishes. This challenged the growers given the change to the thin-skinned Cavendish varieties after the Gros Michel type had been destroyed by the Panama fungus. But consumers insisted—and still insist—upon fruit of roughly uniform length, without visible bruises and blemishes, and certainly not the short, stubby ones, even if they tend to be sweeter. United Fruit invested in overhead cable systems to transport stems from the field and corrugated boxes with compartments for individual hands (clusters or bunches). They invested in pesticide and fertilizers. The extra measure of wrapping stems with plastic while still on the banana plant prevented insect-inflicted blemishes and scarring caused by the plant's own leaves. The independent growers of Jamaica were caught in a world market of colonial pressures and high technology. The only advantage some of them had was cheap labor because peasants had been pushed off land that was good for plantations and had few other job opportunities.[1]

Bananas are the fourth most important crop worldwide for poorer countries, where they provide an important starch source, especially in Africa and Asia. In some African nations as much as 400 kg of plantain, a cousin of the banana, are consumed per year as a main source of calories. The fruit is available year round and provides key foodstuffs between seasonal harvests of other staple crops. The vast majority of bananas grown today are for consumption by the farmers or the local community, with only 15 percent of their global production grown for export. Yet the international pressures may have been far more crucial in the evolution and meaning of the banana. India is the leading producer of bananas worldwide, but so are Guadeloupe, Martinique, Ecuador, Costa Rica, the Philippines, and Colombia. Export bananas go to markets in the United States and Europe, where the "Cavendish" banana prevails. In general, the United States consumes fruits from Central and South America, whereas consumers in the European Union receive most of their bananas from the Caribbean. Ethylene gas ripens the fruit to maturity during the distant but rapid journey to market.

The banana's development into a major worldwide trade commodity has its roots in the nineteenth century and played out after emancipation of slaves, when plantations owners sought other ways—including violence—to tie recently freed small farmers to their production. Individual merchants had shipped plantains from the Caribbean to American and European markets in the early 1800s. But local markets were the center of commerce for the early banana trade. Bananas were produced on small farms by indigenous people;

those bananas not immediately consumed by the farmers were sold in local markets to other members of the community. Visiting merchants thus first gained access to bananas through these local marketplaces and shipped small bunches to overseas markets; thus began the banana's journey of global production, trade, and consumption. In 1804, plantains reached New York and were sold as novelty fruit to curious consumers. Yet despite its entrance into the global market, the banana was not to become a major factor until the turn of the twentieth century, with the rise of steam ships and refrigeration and the expansion of the monoculture plantation under the leadership of such multinational corporations as United Fruit Company.

When I first visited Jamaica in 2008, I was intrigued by its banana, sugarcane, and coffee production, and by the small farmers struggling with local markets and the apparent dominance of international markets for coffee held by white landowners. I visited farms and plantations. I discovered that Jamaica was also one of the leading suppliers of bauxite used to make aluminum in the world. How did a country known for Bob Marley and Usain Bolt, for wealthy white tourists amidst very poor peasants, and for its coffee and marijuana, also develop bauxite? And what significant role did slavery, colonialism, and post-colonialism play in bananas and bauxite? Slavery, sugar, bananas, and aluminum are tied together by the persistence of colonial relations as represented in the aluminum can.

Slavery before Bananas and Bauxite

The modern history of the Caribbean islands is tied to their establishment as jewels of colonial empires in the New World. They served for the extraction of gold and other minerals, and of sugar, tobacco, bananas, and other products. They were subjugated by military force and by technologies that were increasingly efficient at extracting, raising, processing, picking, packing, and transporting the local wealth to the seats of the empires—Spain, Portugal, France, England, and the United States. The local people, in Jamaica the Arawaks, were decimated by Spanish military conquest after 1494 and the disease that accompanied it. The Spanish brought African slaves to replace the Arawaks, many of whom escaped to the hills where, as the Maroons, they later carried out guerrilla warfare against the British. The British brought sugar cultivation and the plantation system to Jamaica along with more slaves, many of whom were exported to other colonies. Hundreds of thousands of slaves provided most of the labor until the mid-nineteenth century; their owners considered them no more than pieces of technology. Katherine Norris writes that 200,000 slaves were re-exported and 800,000 remained to replenish the island's plantation labor force. By the 1833 emancipation, only 320,000 slaves remained, "a figure which indicates that conditions did not favour a long life for an African."[2] White slave owners exploited oppressed African slave women sexually, too, so that by 1820 the number of mulattoes outnumbered whites. Their descendants, and some Chinese and Indians, formed the core of the middle class that sought to join or be part of the British "plantocracy." The plantocracy thrived on inequality; Horace Campbell notes that although there was no legal segregation in Jamaica as in the United States (which the Supreme Court justified in its 1896 decision in *Plessy v. Ferguson*), racism persisted in cultural, economic, and many other institutions, for example, European standards of beauty and status.[3]

The difficult and morally fraught assessment of the relationship between technology and slavery reveals a series of conundrums—and disagreements among specialists—about just how technological change played out in slave societies. Economic historians including those with a Marxist perspective, apologists of the era, and other observers have debated whether the presence of unpaid labor left slave owners without an incentive to innovate, if plantations offered economies of scale that made slavery more prominent than among smaller farmers, and to what extent slaves influenced technological change over several hundred years of servitude. In Jamaica, while our concern is primarily from the late nineteenth century and slavery was abolished in 1833, slavery's impact on social life, work conditions, the economy, the relationship between landowners and laborers—and bananas—continued well into the late colonial period.

Slavery and technology were intimately related. First of all, three-masted caravels enabled ship captains to tack efficiently and cut distances in difficult winds while transporting millions of African people to agricultural enterprises in the New World to toil at plantations and in mines; Europeans used textiles, metals, and firearms to buy humans from Arabic slave traders, and guns were used to enforce servitude. Population densities in the islands remained low. Free laborers had little incentive to migrate for labor, and those freemen who came found cheap land and wished to work for themselves. Native Americans, decimated by European diseases, had no interest in working for settlers, but wished to continue their hunting and gathering, pastoral, or agricultural ways of life. Slavery provided the solution to the problems of agricultural labor in the Caribbean, South America, and the American South, with slaves who worked on tobacco, sugar, cotton, banana, and other farms, and in mining operations, and who therefore produced much of the wealth in the New World.[4] Slaves were beaten, abused, sexually assaulted, kept in chains, forced to work in chain gangs, and murdered.

Slaves were also masters of technology, for example in mining enterprises. The discovery of placer gold deposits in the future Minas Gerais, Brazil, in the 1690s led to a gold rush until the mid-eighteenth century, although the mines were in decline by mid- to late century. Income was uncertain, so investment was risky. Problems included drought, flood, the collapse of a shaft or rock-face, and panned-out deposits. Mining technology was rudimentary because the Spanish Crown refused to employ specialists from Europe whose knowledge of the extent of gold and how to remove it might lead hostile powers to invade. The Crown blocked the development of manufacture in the colonies. Workers and slaves used picks, shovels, and axes, while iron and gunpowder also had to be imported. They mined by panning, pushing ore through gravel beds under hydraulic pressure, and finally filtrating sludge through sluice boxes. Mines therefore needed to employ such skilled slaves as carpenters, masons, and smiths.[5]

Gold mining imposed severe physical demands on slaves, who tended to be young adult males at their peak of physical strength and who were bought on credit from Rio de Janeiro merchants in the thousands. (There were few female slaves, or white females, a situation that led to concubinage.) For panning, slaves often waded waist deep into icy water, while otherwise facing extreme heat. The slaves dug gravel and carried it to a water source. Many of them succumbed to sun poisoning, and also to vomiting and fever chills, acute dysentery, and kidney diseases; pleurisy and pneumonia; and malaria. Accidents, maimings, and death occurred frequently. In the early 1730s, 6,000 whites and blacks died in the Carlos Marinho mines. Slaves were also extremely poorly fed—given rotten food, uncooked manioc, and other dangerous sustenance.[6]

In Jamaica, agriculture was the central activity, especially at first sugar. Slaves were forced to cut and process sugarcane. The sugar industry had its origin in the early sixteenth century under Spanish colonial rule when cane was shipped from Haiti to the island. Initially cultivation was for local use; the Spaniards were unwilling to work on the sugar estates after the Arawaks on the islands were decimated and cane lands fell into disuse. But after the British took control in 1655, they began to develop plantation systems with slaves, as Britain became the major producer and leading exporter of sugar in the world. By the 1770s some 775 sugar establishments had opened. Sugar output reached

101,194 tons in 1805. In the nineteenth century, the sugar industry changed from cheap slave labor of Africans to indentured labor of Indians and Chinese and ultimately to less labor-intensive and more capital-intensive methods. Slaves in the new world were a technology of sugar, cotton, fruit, and other plantations; they were a cheap tool to change the land and to pick crops.[7]

A series of technological innovations accompanied the rise of Jamaican sugar during the period 1760–1830. According to Veront Satchell, the field was a factory with such inventions as the vertical three-roller crushing mill that were essential in Jamaica, yet were nonexistent on farms in the British Isles. He notes that between 1760 and 1830, the Jamaican legislature passed forty-nine bills granting patents for improved methods in sugar and rum production. Of these, thirty-four were for innovations in the infernal sugar-crushing mills, the most important of which were for the application of steam power to the sugar mill. They included an invention to prevent workers' hands from getting caught in the crushing mill. The so-called dumbturner was a circular screen attached to the upper and lower frames of the roller that fed the cane into the crushing mills. Many of the innovations were from artisans and slaves themselves. Satchell writes, "The slaves actively participated in inventing new techniques and equipment pertinent to the sugar industry." Slaves also were blacksmiths, millwrights, coopers, wheelwrights, masons, plumbers, carpenters, coppersmiths, and engineers, and many of them had brought sophisticated metallurgical knowledge from Africa. They manufactured and repaired machinery, arms, and ammunition.[8]

Some individuals argue that the efficiency of the slave trade itself made manpower so readily available that the plantation owners "had little motivation to introduce innovations, or that the plantation owners did not believe their slaves were capable of working with sophisticated tools and thus avoided using complicated machinery."[9] In this environment, how and why might owners strive to modernize mills to increase speed, improve yield, and smooth production? As plantations grew in size, production lagged, and hours of work and tons of sugar might be lost if molasses failed to crystallize. The Spanish, French, and British slowly improved grinding, boiling, purging, bleaching, and drying. They sought to bring science, engineering, agronomy, and modern manufacture arts to the process, to create schools of chemistry and educate new experts. But sugar masters still oversaw the industry and the chemical processes were not fully understood.[10]

By the late eighteenth century high sugar prices and trade liberalization led in Cuba to pressures for new technologies to make sugar mills more efficient. In the absence of modern-day engineers, people on the ground wondered how to apply science and technology and whether to bring foreign technology to the mill. Or, they could increase capacity by buying more slaves to work relentlessly while scaling up traditional production methods. This led to pressure to acquire more land to supply the cane. Laborers cut the cane by hand and transported it by oxcart to the mill. The roller mill, driven by oxen, crushed the cane to extract its juice. This molasses was refined in a boiling house that used kettles of diminishing sizes. Impurities were skimmed off and the concentrate was dried in a curing house. Speed was of the essence since cane rapidly loses sugar content. In any event, a large number of field hands were needed.[11]

Did slave society stimulate inventive activities? Usually innovation occurs when it lowers wages in a firm; if technological change makes production less, not more, complicated; and if it requires less, not more, skilled labor per unit output. Resistance to change often comes from those with skills who are pushed aside by an invention, for example handicraft laborers whose livelihoods are destroyed by mills or factory workers on the assembly line made redundant by machines. But the slave "has nowhere to fall." Aufhauser noted that the association of slavery with technological lag originated with the nineteenth-century antislavery movement, where it was picked up by Marxist historians. He found that opposition to technological change and innovation on eighteenth-century Jamaican plantations did not come from owners or from slaves—the latter having little to lose—but from the skilled sugar masters who oversaw the refining process and were in charge of boiling house operations. Threatened by a science-based production process in sugar refining, the artisans "refused to surrender the command they had traditionally held."[12]

Sugar plantations did not use the plow that could have saved labor in holing the cane either in Barbados or in Jamaica before emancipation, but not because slaves could not use them. Had they used plows, they would become redundant, and with no alternative use the owners faced the problem of letting the slaves starve or feeding them although they performed no labor. They needed great incentives to substitute for labor since redundant labor raised the specter of rebellion or free time. Most writers agree that slaves could use distillation technologies well, but owners usually hired one white man to supervise distillation and clarification because they trusted him to get the work done quickly.

They also preferred to use slaves to cut lumber to feed the hungry processing facilities.[13] Jamaican planters complained about the complexity of new, more efficient machines, but generally employed new apparatuses after the slaves were emancipated.[14]

Although slavery facilitated the growing British sweet tooth, its existence provoked outrage among many citizens. To the morally astute, the slave was a human being. For example, the poet William Cowper joined associates in the Committee for the Abolition of the Slave Trade and penned several abolitionist poems. In "The Negro's Complaint" (1788), he showed the slave not as a victimized object, but as a person with agency.[15]

> Why did all-creating Nature
> Make the plant for which we toil?
> Sighs must fan it, Tears must water,
> Sweat of ours must dress the soil.
> Think ye Masters, iron-hearted,
> Lolling at your jovial Boards,
> Think how many Backs have smarted
> For the Sweets your Cane affords![16]

In the 1830s Parliament freed Jamaica's slaves, who became independent peasants. While their situation improved over the next decades when they cultivated crops other than sugar, the Crown passed new laws intended to help large landowners and force peasants back to the estates as cheap laborers, where they were abused again. The landowners' fields needed hands, while the peasant raised farm animals and tended his gardens. Without sufficient laborers, even after securing indentured East Asians, sugar production fell from 67,850 tons in 1833 to 28,750 tons in 1850, at the same time as the number of freeholders in Jamaica increased from some 2,000 to 50,000.[17] In the assault on free men, the authorities removed 1,200 "squatters" from about 30,000 acres of mostly Crown-ruled and unclaimed lands during the 1870s. Many of those evicted were not squatters at all but merely individuals who, having made purchases of land, did not possess the necessary documentation to satisfy government officials. In this and other ways, as Satchell argues, state policy was responsible for the survival and reconstruction of the plantation economy long after emancipation.[18]

After emancipation, labor remained racist, as white laborers rarely worked the fields; black laborers were linked to white landowners and were exploited both at work and by the laws of the British Crown. Blacks, including the Maroons, often rose up against their exploiters. The Rebellion of Morant Bay in October 1865 indicated the level of frustration among black Jamaicans. They were fed up with their mistreatment, their inability to vote, the damage to their lands by floods, and the failure of government to respond to disasters (see chapter 5) and to such diseases as cholera and smallpox. Some blacks believed the whites intended to restore slavery. The governor of Jamaica, Edward John Eyre, lied in his reports to Queen Victoria about the miserable conditions of blacks; the queen took his word that they were well off, and she told the poor blacks to work harder and be silent. At a trial of a black man accused of trespassing on an abandoned plantation, black Jamaicans could no longer tolerate the injustices. They harassed and threw stones at police. The local militia fired and killed seven protesters. Eyre then established martial law and ordered harsh measures against black Jamaicans. Government troops hunted down loosely organized rebels, who offered little resistance; troops killed 439 of them including women and children and executed 354 more with peremptory trials. Six hundred more women and children were flogged. Eyre ordered the trial and hanging in just two days of George William Gordon, a former slave and politician who was accused of leading the rebellion.

In response to uprisings, the British government placed the island under complete control of the colonial office to ensure the domination of the island's economy and politics by a white oligarchy. As in India, the goal was to extract resources and maintain the empire; at the same time, science, medicine, and technology variously assisted colonialists in the practice of control and rule.[19] The government enabled the immigration of Chinese and Indian indentured immigrants, who worked for low wages, to fight the labor shortages through tax incentives. The justification for this incentive came from the assertion that "negro labor is lazy and will not work." But peasants obviously worked hard on their own lands, and they preferred almost anything to the dangerous conditions and the mistreatment of bosses in cane fields. The rise of the banana and relative decline of sugar led to the dislocation of the peasantry, many of whom emigrated—nearly 150,000 people between the 1880s and 1920s.[20]

Not only labor but especially land figured mightily in the rise of the plantocracy. Government tax policies encouraged the buying up of large swaths of land to receive a discounted rate for owning more than ten acres. Peasants paid eight times as much per acre as the largest landowners because of this system, and also paid taxes on such property as livestock, outbuildings, and houses.

This led to consolidation of land among the wealthy and forced others to rent and work as tenants. Plantation owners used the best lands for their own production and rented out sloped and poorer quality land;[21] this is reminiscent of the way noblemen took the best lands in Russia after the Emancipation of the serfs. To make the most money from a plot of land, a peasant might want to grow a crop like cocoa or coffee, whose trees required years of growth before they bore fruit. Yet frequently landowners evicted the lessee just before harvest, so the landowner would take advantage of all the completed work. Growing a crop likes bananas made sense to avoid a much larger loss on a slower growing crop. Yet this created another problem. The peasants likely leased land on a hillside where it might be vulnerable to runoff. The best crops to grow in this situation would be coffee or some other with roots. Bananas were the most difficult because the plants had shallow roots and the fields required weeding.[22] To make matters worse, poor job prospects for men often forced them to walk six to twelve miles for a day job. Women also labored, and it was hard labor: they collected and smashed rocks for macadamized public roads and they slashed and bundled sugarcane. Miniscule wages prevailed because of the declining sugar industry. Like the Pullman Railway company town in Chicago, Illinois, or many other such company towns throughout the world, price fixing and deductions from pay for food and housing provisions, and payment in scrip contributed to impoverishment. When peasants refused to work, landowners responded by destroying laborers' own crops and houses as punishment.[23] A variety of race and labor issues thus played out in the tension between the plantocracy and the slaves, freed laborers, indentured servants, and peasants.

From 1870 to 1900 small-scale cultivators attempted to enter expanding North Atlantic markets for tropical fruits and such beverages as rum. Their entry to the fruit trade was an effort to live dignified lives and avoid "coercive labor practices and unstable tenancies."[24] They were capable farmers, producers, and transporters; Jamaican women were vital to fruit production as middle brokers. But they faced racism, exclusion from politics, land shortages, devastating hurricanes every five or six years, frequent blowdowns, and determined commodity brokers and shippers who imposed fruit quality and price standards on the bananas that were difficult to meet. When larger corporations bought abandoned sugar estates and converted them to bananas, they forced the price of land up and small settlers could at best afford leases or had to move to the interior. Yet, ironically, these smaller farmers produced the vast majority of food for both local and export markets, including yams, coffee, ginger, fruit, and pimento.

In fact, land has been the central focus of conflicts in Jamaica since the emancipation that centered on sugar, bananas, and later bauxite production. Peasants have naturally wanted more and better land, while members of the plantocracy wished to control it and exploit the peasantry for higher profits, along the way adding racist notions to their exploitative practices. The state sided with the wealthy landowners, and when the bauxite industry arose in the 1940s it sided with the multinational corporations against the peasantry to ensure access to and control over rural land resources. The distribution of land remained unequal into the twentieth century. In the 1950s, farms with fewer than five acres of land accounted for approximately four-fifths of all farms, but together controlled 157,363 acres, or approximately one-twelfth of the total farm acreage, while large plantations exceeding 500 acres in size that accounted for a mere 0.35 percent of the total number of farms controlled nearly three-fifths of the total farm acreage.[25] The banana industry grew in several countries in this environment of struggle between powerful landowners and local peasants, and added to the mix international companies with good political connections.

After Jamaican independence in 1962, sugar production reached its all-time high of 514,825 tons in 1965, but this declined to 186,133 tons in 1998 and to 204,188 tons in 1999, while land under cane cultivation fell from 58,936 hectares in 1973 to about 42,000 hectares in 1999. The cane is processed in eight factories that have become decrepit. Recently, sugar production has fluctuated greatly, between 120,000 tons and 170,000 tons. While declining in share of income, it remains the largest single employer and industry in the agricultural sector, with some 12,000 independent cane farmers who produce just above 50 percent of the cane.[26] Flour (at roughly 130,000 tons annually), molasses (at 65,000 tons), rum (at 26,000,000 liters), and beer and stout (at 85,000,000 liters) are other major Jamaican products.[27]

Bananas and Bugs

Bananas do best at or near sea level. Plantation construction involved clearing large areas of land, rapidly planting trees, and caring for them. A root sends up many "suckers" each of which may eventually mature. Bananas grow well in alluvial soils when planted 4 to 5 meters apart, and if needed with irrigation because bananas grow best with abundant rain and damp soil. Once plants mature, after about eight months, fruit can be gathered weekly. The trees will grow to 2.5 meters in height. "Bunches grow at the juncture of the trunk and branches, and consists of from four to twelve of what are termed hands, each hand having eight to twelve bananas on it," although some poorer plants have bunches with fewer hands. Growers like bananas because they require little skill or labor.[28]

A kind of "banana determinism" prevailed among mid-century observers. They saw the banana as an opening of Caribbean countries to development, ensuring their economic future, building democracy, and overcoming the sparse population "except by poison snakes, ferocious animals, myriads of insects, and dreaded diseases."[29] The lands were backwards, without manufacture, and torn by strife. But New Englanders created the banana industry to bring progress to the islands and money to their pockets. Wrote one geographer, "American engineers are invading the jungles with steam shovels. Swamps are being drained and axes are heard ringing in the woodland. Fruitful banana plantations are appearing as if by magic. The Caribbean lowlands, as they appear today, are a place of prosperity as one-quarter million acres of the most fertile lands have been reclaimed for the use of man, and its sanitation completed." They built 2,000 miles of railroad, connected the banana to the world by radio, telegraph, and telephone. One hundred and fifty modern refrigerator steamships carried passengers, mail, and general cargo to the "banana ports of the Caribbean, and bananas to the United States and Europe." This was a benefit to 100,000 people in the industry; the banana helped them overcome poverty.[30]

The plantations also included living quarters and offices, hospitals, railroads, and docks, and the modern communications technologies of telephone and radio to ensure timing of harvests and transport to railroad and port for quick loading.[31] The capital imported to Caribbean countries from the United States built highways, railroads, public projects, and electric light and water systems. The banana fruit deflected men who might "otherwise be willing recruits for an insurrection . . . to enjoy the fruits of peace and prosperity" of work on plantations and promoted "a friendlier feeling toward the United States."[32] Like the Tennessee Valley Authority, whose electricity would bring democracy to the hollows of Appalachia, so the technology of the plantation would secure Caribbean nations from the lack of civilization.[33]

The Gros Michel variety of banana reached Jamaica in 1835, when botanist Jean Francois Pouyat planted a single specimen from Martinique. It spread rapidly by Afro-Jamaican farmers, and existed only for a generation or two before export trade began.[34] Several North American shippers began buying fruit in the late 1860s using schooners and steamships to transport them; they bought from wholesale dealers along Jamaica's northeast coastline. They liked the Gros Michel because of its relatively thick skin and its small bunches that traveled well when tightly packed. Wind-powered vessels dominated the trade, accounting for two-thirds of exports from Jamaica as late as 1878, and, in part with colonial government subsidies, steamships supplanted them within a handful of years. Shippers set strict standards for when and where to sell their fruit so that they got a steady supply of bananas that would not ripen prior to reaching markets.[35] Jamaica was the source of the Gros Michel for nearly the entire globe. Fiji (1891), Colombia (1892), Hawaii (1903), Surinam (1904), and Australia (1910) all received "suckers" from Jamaica.[36]

In 1804, Cuban shippers sent thirty bunches of bananas to New York, and by 1830, 1,500 bunches. By 1857 regular trade existed between Cuba and Boston, and in 1869 bananas began their journey from Port Antonio, Jamaica. Many small farmers continued to raise bananas and sold their bunches to market miles away. Plantations solved this problem. If in 1879 only one estate near Port Antonio was identified as a "banana plantation," then by the early 1890s there were more than one hundred banana plantations. Panama joined the banana trade next, although in the 1880s it virtually ceased due to canal construction. Costa Rica and Jamaica filled the banana gap. The Boston Fruit Company grew out of increased banana trade, itself spurred by the replacement of schooners by steamships for carrying bananas; steamers could carry 30,000–60,000 stems, three to five times that of schooners. The bananas were kept cool by ventilation cowls, which gave way to refrigeration with forced cold air by the early twentieth century. They sailed to the United States in less than a week and to Britain in somewhat longer than two weeks. US imports increased from almost $490,000 in 1875 to $2.1 million in 1885 to $4.7 million in 1890 to $6.6 million in 1901. Producers established plantations to ensure full, regular cargo and create a steady demand at a fixed price. In 1884, Boston's Andrew Preston, who sold Jamaican bananas, joined with nine other business men to monopolize the banana business with Boston as the port of entry.[37]

By 1902 in Jamaica the United Fruit Company (formed in 1899 following the merger of the Boston Fruit Company with twelve other firms, and today the Chiquita International Brands) cultivated nearly 8,000 acres of bananas—more than one-fourth of the banana acreage in the four major exporting parishes, using roughly 600 indentured South Asians and black Jamaicans. Approximately 2,700 South Asian men and women arrived in Jamaica between 1899 and 1906 to work on banana farms. They were subsidized by the government, which provided health care. The plantations' owners felt the workers were needed since they could not rely on small holders for labor as they had their own farms and concerns. By the beginning of the twentieth century, small holders declined in absolute numbers and in total banana production because of falling fruit prices, increasing land prices, and occasionally the refusal of the big companies to buy their fruit.[38] This forced small holders to lease land or seek plots at higher elevations that were not good for bananas, so many of them switched to cacao, coffee, and ground provisions.

To modernize the banana, cacao, citrus, and sugarcane production, the British established the Imperial College of Tropical Agriculture in 1922. The college built on work conducted in other colonial and domestic institutions that was intended to bring science to the colonies and products to Britain, based on recommendations of the Colonial Office Committee on "Agricultural Research in the Non-self-governing Dependencies." The college had research and postgraduate studies. Specialists included soil scientists, biologists, entomologists, and mycologists. The goal was "to develop the agricultural possibilities of our Colonies satisfactorily." The researchers believed they faced many investigative problems and had few workers, so they decided to concentrate on those issues that had direct economic bearing. Overestimating the power of science, standing firm in their colonial beliefs, and underestimating or not understanding ecosystem difference, they believed the few graduates they turned out annually could serve anywhere, for there was a similarity in problems and goals "whether it be rubber in Malay, cotton in the Sudan, coconuts in the S. Pacific or sugar-cane in the West Indies."[39] As Michael Adas demonstrates, science and technology served conquest and colonialism in the nineteenth and twentieth centuries by indicating to Europeans that their "mastery" and understandings of nature confirmed their belief in the superiority of the civilization they brought to others when they sought to civilize and control that colonial world.[40]

The close relationship among manufacturers, plantations, and research left the small farmer without the benefit of applied research. For example, the Cacao Research Scheme was financed conjointly by the colonies and the big chocolate manufacturing firms. It was launched in 1930 and consisted of two branches, namely, botanical and chemical, and resulted in the selection of trees with the highest numbers of pods and qualities.[41] At the same time, as with the colonial experience everywhere, officials and specialists viewed the peasants as backward, ignorant, and environmentally destructive. Articles in the *Journal of the Jamaica Agricultural Society* offered paternalistic technical advice on agriculture that reflected racist ideas of white Anglo-supremacy and ignored the contribution of the plantation to the island's social and environmental problems. Yet government officials still promoted export banana cultivation as an "all-Jamaica" project.[42]

Doctors of the British Medical Association, like their counterparts in the USSR who studied every potential food to determine its potential nutritional value for the Soviet worker, evaluated and touted the joys of new foods, especially fruits, because they considered the British diet to be defective. The nation overcame this problem with "a marked increase in the consumption of fruit by all classes," as the use of the banana indicated. At first only the wealthy consumed it, "but now it is universally sold on the humblest barrows and stalls, and forms a not inconsiderable item in the diet of the poor"—not to mention Sandburg's peddler. But the British were not entirely clever: they consumed bananas before they were ripe, and when they were "insipid in flavour, and to cause dyspepsia and other forms of intestinal disturbance. They should not be eaten before the skin is blackened in places, or when there is any reluctance in the skin to separate from the pulp." Children surely knew best to eat the soft, ripe ones. The doctors urged attention so that vendors would be prohibited from selling bananas that had over-ripened.[43]

By the late 1920s nurses touted the banana as treatment for celiac disease, supplements in the diet when treating TB, and for people with nephritis.[44] Professors of industrial microbiology praised the banana for its low cost and high nutrition in comparison with wheat, rye, barley, and potatoes, in food produced per acre, and in salts (electrolytes). Its skin protected the food from dirt and bacteria, even if the skin was 35 percent of total weight, and if bananas contained a bit less protein than potatoes, they were easily digestible and inexpensive.[45] Banana leaves had other functions; in New Guinea girls wore shredded banana leaves as "petticoats" as part of the practices surrounding their first menstrual periods.[46] The presence of warships in all of the oceans interrupted deliveries of bananas during World War II and caused great shortages. This was a problem for some families, since bananas had already become such an accepted and popular baby food. The authorities addressed this issue by urging parents to wean children slowly from bananas to other foods.[47]

By the early twentieth century, diseases struck banana crops and spread rapidly because of monocultural (plantation) approaches. Panama disease (*Fusarium cúbense*), a soil-borne fungus, led to large-scale botanical research to seek banana varieties that were immune or resistant. Botanists, plant breeders, plant pathologists and physiologists, and cytologists worked to find an alternative to the Gros Michel. The banana then fell to Sigatoka disease.[48] Even when diseases caused specific problems, specialists at Kew Gardens had difficulty in determining the source. In one case, Kew reported to the British Colonial Office that a Fiji disease was possibly produced by a nematode worm, and "recommended, failing success with various insecticides, plowing the land, leaving it fallow, and alternating some other crop. The ground could then be re-planted with banana 'stools' from an unaffected locality. Aphids, which seemed not to be the culprit, and a root fungus, that seemed more likely, were found."[49] Researchers undertook 5,000–6,000 pollinations to develop a resistant hybrid, but without success. They held out hope in the 1930s to find a new breed "with the necessary commercial attributes such as seedlessness, long fingers, good flavor and colour, the right shaped bunch for shipping, and so on." Specialists in the Colonial Office, the Royal Botanic Gardens, Kew, and the agricultural departments in other colonies assisted the researchers in finding new species that might substitute and also meet the needs of ripening and transport since "the consumer in England does not realize that the banana he eats is really only about half grown."[50]

As is true of any monoculture, the banana monoculture required—in the minds of owners and managers of plantations—the liberal application of pesticides and other biocides to control pests and diseases in order to produce a lush garden of one crop. The companies ordered workers to apply the chemicals, thereby exposing workers to dangerous substances. For example, in Costa Rica, the United Fruit Company dictated hand spraying—labor-intensive work—of banana plantations to control Sigatoka disease, a fast-moving, airborne fungal epidemic, from 1938 through 1962. (The fungus spread through Honduras, and into Belize, Mexico, and Jamaica.) The workers suffered headaches, night coughs, bad eyesight, and believed they were prone to tuberculosis. This spraying program was one of the first in Central America and remained the largest one in the world. At the same time, United Fruit was fighting the soil-borne fungus Panama disease. Both diseases were "less a natural disaster than a product of industrial-scale, globalized agriculture."[51]

Any monoculture will be prone to epidemic disease; the spruce budworm in Maine destroyed vast holdings of the Great Northern Paper Company that clear-cut vast tracts of land, then used herbicides to create spruce forests without hardwoods, then turned to pesticides against the worms.[52] When United Fruit's plantations in Costa Rica reached 50,000 hectares in massive uniform blocks of 300–600 hectares, the bananas were ripe for the picking, but they were also ripe for rapidly spreading disease. Company scientists selected copper sulfate as a fungicide and, after air dusting failed, they decided to deliver the 250 gallons needed per acre, twenty to thirty times per year, by hand. They built a huge system of iron piping, hoses, and pumping plants manned by more than a quarter of the labor force of thousands of workers. The workers carried out this hard and dangerous work, and finished the day with chemical spray and its mixing agents stained on their clothes and skin. Since they had to spray leaves on both sides of trees as high as 40 feet tall, the residue and spray often fell in their faces, eyes, and skin, and for months later their nasal discharge and sweat were green-tinged.[53]

But the big companies would not be deterred. They succeeded in turning bananas into an everyday, inexpensive snack. They found varieties that could grow year-round. They cleared rain forest and built railroads to move the banana. They worked with local "banana republics" to keep prices low, prevent unions, and pay low wages. In fact, in 1954 the US government intervened in Guatemala to overthrow a democratically elected government at the insistence of United Fruit.[54] How absurd that the banana was central to US economic and political policies directed toward keeping klepocratic Western Hemisphere dictators in power through the exploitation of the peasantry. This was the "banana republic."

In 1904 O. Henry (pseudonym of William Porter) published his *Cabbages and Kings*, a book about the fictional Republic of Anchuria, loosely modeled on his experiences in Honduras. He described that tropical landscape without affection: "Spaces here and there had been wrested from the jungle and planted with bananas and cane and orange groves. The rest was a riot of wild vegetation, the home of monkeys, tapirs, jaguars, alligators and prodigious reptiles and insects. Where no road was cut a serpent could scarcely make its way through the tangle of vines and creepers. Across the treacherous mangrove swamps few things without wings could safely pass."[55] O. Henry playfully noted that "in the constitution of this small, maritime banana republic was a forgotten section that provided for the maintenance of a navy. . . . Anchuria had no navy and had no use for one. . . . With delightful mock seriousness the Minister of War proposed the creation of a navy. He argued its need and the glories it might achieve with such gay and witty zeal that the travesty overcame with its humour even the swart dignity of President Losada himself."[56] Hence a banana republic is a dictatorship built upon corruption, enrichment of the wealthy landowners and generals, the exploitation of the masses who toil on plantations, and the pursuit of military hardware at the expense of the public good.

In the period 1914–1930 banana companies diversified, but faced extensive competition and great losses from floods, droughts, blowdowns, and banana disease. Irrigation, research and development, and better transport and organization helped a great deal, but could not make up for epidemic pestilence. Companies diversified into "sugar, and raise coffee, pineapples, and citrus fruits; they [were] awarded valuable mail contracts in connection with their businesses as common carriers; they own and operate railroads, telegraph and radio stations, hotels, banks, mercantile establishments, hospitals (serving a large public as well as their own employees), electric light plants, and engage in many other activities intimately connected with the peace and prosperity of the countries of the Caribbean."[57]

Almost inevitably, the banana industry declined from World War I, with the United Fruit Company abandoning lands and clearing new land. Without constant intervention, without expenditures on capital inputs of chemicals, land, and machines, and even with scientific research, the monoculture of bananas fell, not because of soil exhaustion, but because of blight, especially Panama disease. "Panama disease attacks the plant through the root system, causing the leaves to wilt and the plant to rot off at the ground. An infected plant does not produce a bunch of bananas. Infected soils cannot be used for growing bananas again for years." Specialists found a temporary way to control the spread of the disease, adding hydrated lime, phosphate, and potash to the diseased soil and planting to beans, plowing under the beans for some years, and then hoping to replant bananas. But the losing battle meant the Fruit Company contracted out to independent growers whose smaller farms, still at risk, had not yet succumbed. Then in 1938 Sigatoka leaf spot disease hit, and the Company withdrew and turned to cacao, abaca fiber, and rubber.[58]

In the early 1900s the United Fruit Company set up its Golfito, Costa Rica, division on the southwest Pacific Coast in a sparsely settled tropical rain forest. It built 246 kilometers of railway lines; constructed modern banana shipping facilities, with equipment for loading 4,000 bunches of bananas an hour; built a company town of 7,000 residents with a hospital, stores, construction, communication, recreation, and other modern facilities; and brought into production thousands of acres of bananas. It hired 15,000 white and mestizo workers. It assembled a massive irrigation and pumping system that ran around the clock; the standard pump handled 5,000 gallons of water a minute and fed ten towers. And then Sigatoka hit.[59] Still, by 2012, Costa Rica was exporting more than 1.2 million tons of bananas, although the industry remained under constant threat from the insect and mealybug infestations that spread across 24,000 hectares of land. Panama disease continued its march around the world. Pesticide use has doubled across Central America and in Costa Rica in the twenty-first century and is leaching into the Rio Suerte and thence into the Tortuguero Conservation Area, where it attacks flora and fauna.[60] A crisis is certain to ensue, but for now we still get our bananas in North America and Europe.

Aluminum Cans

In 1938, people across the Caribbean rebelled against economic and racist exploitation under the British Crown. The problem remained as it had been in the eighteenth and nineteenth centuries: a white ruling class that owned most of the land dominated political power. Its representatives considered themselves British and civilized and their source of labor lazy and uncivilized. They sought to control the islands including Jamaica as part of the empire.[61] The Jamaican revolt spread to every parish. The fighters blocked roads, cut telephone wires, destroyed bridges, burned cane, destroyed banana fields, and on several occasions ambushed police with nothing but sticks and stones. Given the location of bauxite reserves within agricultural areas, conflicts continued after World War II with the state often siding with the corporations against small landowners. In addition, the bauxite industry did not require significant numbers of laborers relative to agriculture.[62]

While requiring copious amounts of electricity to produce, aluminum became a wonder metal during and after World War II for its flexibility and light weight. World production of aluminum rose from 8 million tons in 1950 to more than 25 million tons in 1960. Production commercially dates to the 1890s and doubled about every ten years in the second half of the twentieth century. Roughly 90 percent of bauxite is used for alumina and 90 percent of that is used in aluminum. The remaining 10 percent is used in chemicals, abrasives, refractories, and insulators. Bauxite is also used directly in the manufacture of cement, oil filtration, and in iron smelting. Jamaica produced 2.7 million tons of bauxite in 1950 and 5.3 million in 1955; most of this product went to multinational corporations in the United States and Canada. The major producers of aluminum in 1965 were the United States (2.4 million tons), Canada (1.1 million tons), and the Soviet Union and Eastern Europe (2.3 million tons).[63] Aluminum was a Cold War necessity, often produced by smelters near massive hydroelectric power stations that produced electricity relatively cheaply—on the Columbia River, the Tennessee Valley rivers, Volga River, the Siberian rivers.

Americans consume almost twice as many soft drinks as any other country and 216 liters annually per capita. Roughly 50 percent of Americans drink a soft drink daily. According to the National Soft Drink Association, which defends the rights of Americans to drink sugared, flavored water, Americans consume 600 twelve-ounce servings per person per year. According to the Centers for Disease Control, males consume many more sugar drinks than females; teenagers and young adults consume more sugar drinks than other age groups; approximately one-half of the U.S. population consumes sugar drinks on any given day; non-Hispanic black children and adolescents consume more sugar drinks in relation to their overall diet than their Mexican American counterparts; and low-income persons consume more sugar drinks in relation to their overall diet than those with higher income. And many of them do so from aluminum cans.[64] (See chapter 4 on corn sweeteners and potatoes.) Worldwide production is nearing 500 billion cans per year worldwide. Plastic bottles are equally over-used. Americans use 160 to 170 bottles each year on average, rarely recycle them, and use 50 billion yearly in all.[65]

Bauxite is located primarily in the parishes of Clarendon, St. Catherine, St. Ann, Manchester, Trelawny, and St. Elizabeth in the center of the Jamaican landmass. In the 1940s North American mining enterprises descended upon these parishes in the central limestone regions of the island where, until this time, export agriculture, pastoral activities, and small-scale farms prevailed. Jamaican bauxite was attractive to US companies since it was so close to the US mainland when military demand was growing exponentially to build fighter jets, bombers, and other military materiel. The Jamaican government, part of the British Commonwealth, passed an emergency law in 1942 that nationalized all lands with bauxite deposits. This legislation generated further debate about land rights, leading to the Minerals (Vesting) Act and the Mining Act, both in 1947, that erased any uncertainty. Three companies,

Reynolds (Jamaica) Mines, Kaiser Bauxite Company, and Jamaica Bauxite Limited (whose name was later changed to Alumina Jamaica Ltd. in 1952 and Alean Jamaica Ltd. in 1962), had already acquired vast tracts of land (between 1943 and 1957 a total of nearly 140,000 acres or approximately 5.7 percent of the island's land area). They displaced large and medium animal operations as well as peasants. By the mid-1950s virtually all of the leading grazing areas in the six parishes had come into aluminum's hands.[66]

A major facet of technological change is the phenomenon of the ousting of large numbers of people from ways of life and traditional homelands. Dislocation happens in the processes of migration and urbanization, although these two terms do not reveal the extent of disruption of life that occurs. In many cases, moving away from one life to another happens almost seamlessly, and may be almost unnoticeable, for example during the first stages of the Industrial Revolution when craftsmen, women, and farmers and their children moved to burgeoning towns and cities. Many of them supplied new manufacturers with labor, a few in number went to relatively small factories, and then more and more of them ended up in larger mills, a process that came to feel violent as they were uprooted, and this provoked a violent response—strikes, rebellions, smashing of machines by the Luddites in the early 1800s. Eventually the world changed from pastoral to industrial and from small-scale agriculture to larger farms and plantations based on monocultures. In the twentieth century, as part of the evolution of large-scale technological systems, "ousting" took place not simply as the unfolding of larger, seemingly invisible socioeconomic trends, but as part of massive state plans that were based on the certainty of "progress." Although not one planner could actually define progress, it frequently resounded with the promise that relocated people would find newer, better housing, modern conveniences, better land, and good social services that would mean a better life.[67] But whether referring to the farmers removed from floodplains of Brazilian rivers to new, irrigated fruit farms, or the 1.5 million Chinese peasants ripped from traditional farms and homes and forced to leave cemeteries and schools behind as the Three Gorges Dam inundated their lives, or the millions of other people displaced by dams, highways, and reclamation projects throughout the world, the results are mixed at best, and more often than not it is the poor, powerless, underclass, the peasant, the person of color who bears the brunt of technological change as an "oustee."

This was no different in Jamaica. Bauxite operations commenced only in 1952, but farmers had been pushed aside already to enable mining. From 1953 the bauxite industry expanded rapidly. It produced 340,420 net dry tons of bauxite in that year, 5,745,002 tons in 1960, 11,820,076 tons in 1970, and 13,385,668 tons in 1973. Alumina production expanded from 100,471 tons in 1952 to 1,595,467 in 1960, 4,244,590 in 1970, and 6,112,089 tons in 1973.[68] Jamaica's bauxite production increased rapidly in response to Cold War stimuli connected with Canadian and American production. The deposits are excellent for opencast mining because of overburden limited to 2 feet; specialists estimate reserves at between 350 and 400 million tons. Investment by the three bauxite companies up to 1959 exceeded £40,000,000; at that time the companies employed more than 6,000 people,[69] far fewer than the companies had promised when securing land and tax benefits, and not in the least making a dent in endemic unemployment.

By 1957, Jamaica had become the largest producer of bauxite in the world and the main supplier of bauxite to the North American market. Other companies entered the market: Alcoa Minerals (Jamaica) Ltd. in 1959, Alumina Partners of Jamaica (Alpart), a consortium of Kaiser and Anaconda in 1969, and Revere Copper and Brass in 1971. They also acquired lands, although it was difficult to buy large tracts earlier available, and thus encroached more and more on small peasant holdings. They displaced many people, although by law the companies were required to provide for resettlement.[70] After independence in 1962, the call for nationalization of the bauxite-alumina industry grew loud, at the same time as public opinion came to see that promised "public good" of jobs had not materialized in bauxite, but instead required taking care of the nation's underemployed agricultural workers. The companies became "intruders," and many peasants refused to sell and move out. Still, the companies continued to acquire land, but in smaller plots, often under 25 acres or less, especially in central Clarendon and northern St. Elizabeth parishes.[71]

Besides the fact that they did not help in resettlement nor provide the promised jobs, another problem associated with mining was environmental degradation. Having acquired vast portions of land on the cheap, the companies diversified somewhat with forestry, cattle, and other agricultural operations. The forestry operations were a paradoxical addition to their efforts since the aluminum companies destroyed land, soil, and forests through bauxite mining. The forest supervisor of Alean Jamaica Limited noted that the mountains of the parishes of Manchester and St. Ann were far too steep and wooded for agriculture, with little topsoil and little organic material, with low moisture-holding capacity and undulating or mountainous terrain with relatively narrow valleys between limestone hills. Peasants did their best to grow crops in the mountains. But when Alean first acquired lands in 1943 to 1947, much of it was deforested and eroded by small farmers who, although they had ingeniously adopted small-scale practices to slow erosion and worked very little land in total, were considered responsible for deforestation, while the massive bauxite operations that scarred the earth were considered paragons of stewardship. The newly acquired lands were closed to cultivation and the tenants were moved when possible, which allowed natural regeneration of shrubs and trees. The company followed with reforestation, using pine, hibiscus, mahogany, and fiddlewood. Hardwoods worked better than pines in limey soils. In the early 1970s the company planted trial plots at an elevation of 1,200 to 1,400 feet and with average rainfall of 70 inches annually and determined that the Bahamensis variety of *Pinus caribaea* might work.[72] Reynolds in the 1950s planted some 400,000 hardwood seedlings and Alean some 300,000 seedlings to establish approximately 300 acres of forest lands. Alean had more than 5,000 head of cattle by 1962, while Reynolds had a herd of 16,000 head by 1960. Livestock production intensified in the 1960s and 1970s. The companies also leased approximately 50,000 acres of land to tenant farmers who produced yams, sweet potatoes, Irish potatoes, corn, peas and beans, with very little tree-crop activity.[73] As far as can be determined, the afforested areas were neither extensive nor carefully tended.

Jamaica pursued the restoration of abandoned strip mines based on laws passed in 1947–1950 against long odds. Initially local political leaders assumed that mines simply made human habitation within 20 miles impossible, let alone enabled sugar, coffee, bananas, coconuts, fruits, vegetables, and livestock production. The laws required that the mining companies maintain an agricultural operation immediately after buying or leasing land; the basic idea was that for every acre mined, another acre had to be put into agricultural use. Yet Jamaica remains a food-deficient nation because most land is of marginal fertility or in mountainous areas, and the land best suited for agriculture once again has often fallen into corporate hands. Poverty resulted for many small farmers.[74] Reynolds tried to grow sugar on its restored pits but the land was so infertile that the harvest was worthless. Grazing was more successful. From 1952 to 1970, Jamaica restored 3,700 acres of abandoned mine pits, an average of 205 acres per year. The cost of restoration was very high, at thousands of dollars per acre.[75]

In the 1970s the government of Prime Minister Michael Manley pursued economic reforms in the name of the people and socialism. One project centered on reducing high unemployment through creation of agricultural cooperatives, land redistribution, and other programs. Another involved nationalization of foreign-owned electricity, telephone, and public transportation companies. The government also increased levies on bauxite that led to conflict with the American and Canadian aluminum companies, while being met with widespread national support and earning the country millions of Jamaican dollars. The Manley government gained larger shares of aluminum companies, plus a large share of the companies' agricultural lands; Revere Aluminum withdrew from Jamaica rather than agree to changes in the law. Eventually the companies developed bauxite in Brazil, Australia, and Guinea rather than cooperate with the Jamaican government. This led to decline of the industry. In 1974 Jamaica was the world's second largest producer of bauxite and second largest exporter of aluminum, but by the 2000s it had fallen to sixth place or lower, its output had dropped from 18 percent to 7 percent of the world's total, and half of its alumina capacity was closed in 2012. In 2014, about half of Jamaican bauxite production was owned by the Jamaican government, but roughly 90 percent of its alumina production was owned by Rusal, which is based in Moscow, under the direction of Oleg Deripaska, one of the twenty most wealthy people in the world and a member of Vladimir Putin's inner circle.

Technology, Colonialism, and Postcolonialism

In Jamaica, technology mediated difficult relationships between land and people. From conquest to the introduction of slavery, from emancipation to plantation and independence, and from sugar to aluminum, Jamaican people disputed land and labor. Whether as slaves producing cane, peasants pushed aside for banana farms, indentured workers, or small farmers trying to establish healthy relationships with family, friends, markets, and multinational corporations after the appearance of bauxite mining, they struggled against imperial, colonial, and postcolonial relationships established to benefit distant people, representatives of the crown, a new ruling class, and multinationals who sought to keep them poor and off balance. They worked with planting, harvesting, processing, felling, and transportation technologies to ensure the economy worked. And they supplied the natural and mineral resources to provide billions of consumers around the world with aluminum cans filled with soft drinks.

What is Jamaica's future? With over one-half of the country's alumina capacity closed in 2012, and output hovering around 10 million tons annually, Jamaica's position in the world industry continues to slip.[76] According to

Earle V. Roberts of the Sugar Industry Research Institute there may be hope for sugar. In the early 2010s, the Cane Expansion Fund offered loans of $154.3 million to replant of 600 hectares of sugarcane, and there were other loans to expand irrigation. (The government also contributed to human capital through housing and other programs.) Sugar production increased in 2010–2011 by 14 percent and its quality was better. The 2011–2012 crop year was the first since the completion of divestment of government-owned estates which was intended to increase efficiency and productivity.[77] But the vagaries of Jamaican agriculture and decrepit old mills made this hope a dream; the following year the crop was only 132,000 tons, or 5 percent lower than the previous year,[78] and the mills work at only one-half capacity. Still, sugar employs 50,000 workers and has great potential for biomass someday.

Another, perhaps less happy chapter of the banana may have arrived. Oil and transport costs have gone up. Another disease is attacking the Cavendish banana that replaced the Gros Michel.[79] Bananas result from vegetative propagation, not pollinated sex. This means "no annoying seeds, which may be good news for hungry consumers but also means that there is little or no genetic variation—and hence little or no resistance to the banana's many natural enemies." Thus scientists have turned to the banana genome project in the hopes of introducing some genetic variety into the banana and its cousin the plantain. Perhaps they can engineer a banana that will not succumb to Sigatoka or Panama disease.[80]

In his poem, "United Fruit Company," Pablo Neruda, the leftist Chilean writer, attacked such multinational corporations as Coca-Cola, Ford, and Anaconda Copper for parceling out the land, with the Fruit Company reserving "for itself the most succulent, the central coast of my own land." The Company christened "banana republics," praised liberty and flags, and yet established "the dictatorship of the flies, Trujillo flies, Tacho flies, Carias flies, Martines flies, Ubico flies," and other tyrannies in pursuit of coffee and fruit, while the citizens fell into "the sugared chasms of the harbors."[81] It's hard to justify a sweet tooth when one realizes how sugar got to the table, how little it has changed from its labor-intensive foundations in slavery and in exploited workers in the twenty-first century, and how it has destroyed hundreds of thousands of acres of land. But for many people, bananas are a food, not a technology with a story, and so are very hard to give up.

Potato digger near East Grand Forks, Minnesota, early 1940s (?). Like many other foods, potatoes have become industrial products and part of large-scale technological systems. Library of Congress, Prints and Photographs Division, FSA/OWI Collection, LC-DIG-fsa-8a22251.

4

Mass-Produced Nutrition

Industrial Potatoes, Industrial Sweeteners

Some years ago I stumbled upon an article in a Soviet newspaper that proclaimed "Nuclear Chickens!" The author suggested the joys of radiation sterilization of processed chickens to increase shelf life of the meat. Given the paradoxical failings of refrigeration and freezing in snowbound USSR, radioactive isotopes provided a solution to food deliveries without spoilage. Of course, they first focused on potatoes, then other foodstuffs. Since the time I read about those unfortunate birds, I have actively considered how my food gets to my table and what is in it. How can I be sure it is safe and good for me?

I have been an inveterate label-reader since I was 12 years old. My mother will tell you that I could eat an entire box of cereal at one sitting, and often did, carefully reading the packaging many times. I loved Frosted Flakes, perhaps because I did not understand how worthless that cereal was for me. Yet my first encounter as a 13-year-old with FDA data indicated that my favorite cereal was all carbohydrates, little fiber—in fact, very little that was nutritious. The labeling then as now is an exercise in obfuscation because the food industry has captured much of the regulatory impulse and it is in the industry's interests not to reveal quickly and clearly how nutritionally poor most processed foods are. I was also an inveterate newspaper reader and I remember reading a story in the *Pittsburgh Post-Gazette* about an FDA study that my beloved Frosted Flakes ranked something on the order of fifty-seventh out of sixty cereals in nutritional value. As I recall, the author added that it was "Probably better to eat the box." I asked my mother to buy one of the top three cereals instead from that time on, although it too was an adulterated product, with most of the "goodness" first taken out and then added back as bleached this, processed that, plus preservatives, iron, niacin, vitamins, and other supplements not "naturally" present. And "natural flavors," whatever those are.

Reading labels will never be a science in the United States because manufacturers of our food, working closely with members of Congress, created a confusing system for listing and evaluating the contents. There's less glasnost here than in Gorbachev's Russia. You can mostly divine what's in a packaged food, but not how much of the ingredients, only their approximate rank order, and there's almost always much less of the major substance than what's purported to be ("jelly" on the label, for example, rarely means fruit, but mostly concentrate, juice, and high fructose corn syrup). In other words, the labels mislead you on their contents and much of that content is engineered. Food scientists have engineered the stuff in boxes, cans, and packages to ensure our taste and other dependencies on it. Paradoxically, when a food is less engineered, has fewer additives, and is locally produced, it will cost more, even though it will lack such ingredients as emulsifiers, salt, shelf-life enhancers (not radiation, but preservatives), and other chemicals whose real contribution to health is not vital, and it will certainly cost less to transport locally than long distance in those ubiquitous food-service tractor trailers visible everywhere along the Interstate Highway System. While the FDA determined that radiation is not an "additive," it requires irradiated foods to be marked with the Radura symbol, but not so marked as individual ingredients in multi-ingredient foods.[1] I have yet to see the Radura symbol, but I always find a hell of a lot of high fructose corn syrup (HFCS, or "fruc" as I will call it here) in the mix.

US residents eat nearly 150 pounds of sweeteners annually, and increasingly they are in the form of HFCS—especially in processed foods, soft drinks, and sports drinks. They are known also as fruit juice concentrate, corn sweetener, invert sugar. Sugar consumption has increased 30 percent in thirty-five years. Scientists figured out how to convert the glucose in corn starch into fructose, usually at 42 percent or 90 percent, which is then diluted to make 55 percent HFCS. According to other sources, the average US citizen consumes 2,175 pounds of food annually with average daily calorie intake of 3,600 versus the world average of 2,700, and with one-fifth of that from fast food. Fructose alone accounts for 10 percent of total calories in the American diet. Coke serves 1.7 billion servings of Coca-Cola daily according to the company's 2010 annual report. Granted, fruc helps increase shelf life. But so much fruc consumption will not increase your shelf life.

How did we become dependent on fruc? How did we so easily abandon lo-cally grown food for industrial products manufactured far away? Rather than blame the individual for eating too much, the analysis of food production, distribution, and consumption indicates that we eat too much and too many of the wrong things because of a technological system consisting of corporations and crops and engineered foods that make it difficult to find healthy alterna-tives, and because of confusion over food quality and health concerns created by careful food scientists and media specialists. People know advertising works; it worked for cigarettes and other unhealthful products, and thus we must avoid entirely blaming the consumer for the state of affairs.

In fact, this is a story of self-augmenting technologies, industrialization, and globalization of food trade. It is a story of the combination of US Depart-ment of Agriculture research, land grant colleges, federal subsidies, Marcusean trade associations, organizations that speak with an Orwellian lexicon, and the grotesquery of seeking profits at high cost to human health. What is the economic logic of industrial food production? What are its technological roots? We can understand the ubiquity of French fries and sweetened salad dressings, ketchup, stuffing, sweetened ice tea, bread and muffins, sweetened breakfast cereals, even cough syrup, whipped cream, cottage cheese, yogurt, pickles, ice cream, jam and syrup, sandwich meats, snacks, and soups through the lens of history of technology. This is a story of the power of the industrial food estab-lishment, its roots in an unexpected government–private sector partnership and concerns about what constitute "facts" in seemingly scientific debates.

Self-Augmenting Food

Foods have become extensions of a self-augmenting, self-replicating, highly industrial world in which engineering knowledge, scientific research, and capital are melded together to produce monochromatic edibles—often under golden arches, or in red rooms, or in plastic boxes of faux fishermen. I consider these foods, for example potatoes and tomatoes, to be large-scale Ellulian technolo-gies. Jacques Ellul, a French sociologist, published *The Technological Society* in 1964, setting forth one of the most profound and persuasive arguments for the phenomenon of "technological determinism." He argued that technologies act independently of human agency, in fact requiring us to do their bidding as each technology grows and augments itself. Indeed he called it "la technique"—for which he never truly provided a clear definition—self-augmenting and autonomous, seeking only efficiency. My own favorite Ellulian example is the automobile which begat the road, highway, and parkway; traffic signals; gas stations; fast food; and a US foreign policy dedicated to securing oil around the world through drone military technologies.

While discredited by historians of technology as ignoring human agency, the notion of technological determinism still has the power to frighten us when thinking about whether our choices for a variety of technologies and their support systems is limited. These include the closely connected relations between rivers, impoundments on them, hydroelectricity stations, irrigation systems, fruits and vegetables, and . . . fast food like the French fry. The development of the Columbia River basin in Washington State since the late nineteenth century provides an Ellulian example of the growth of potato agribusinesses that supply much of the world's potato product. The massive dams and extensive irrigation systems built in the second half of the twentieth century enabled conversion of fertile but arid volcanic soil in eastern Washington, Oregon, and Idaho into fruit and vegetable farms. Rather than the independent, small farmer for whom the land and water were intended, they have been acquired by agribusinesses that benefit from an irrigation-university research-corporate trade organization system that employs extensive advertising and lobbying to ensure that the potato and other engineered food products have a major place at our table, in our schools, and around the world.

In *The Organic Machine* Richard White examines how humans transformed the Columbia River using machines. He explores ideas about nature, technology, and progress, and he offers a strong answer to the determinist. From boats to steamboats, to weirs and dams, to steam-powered canneries and railroads, to the hydropower stations of the twentieth century, humans sought to free themselves from hard work and eventually transformed the river into an electrical grid, its abundant salmon having all but disappeared because of the dams placed in their way for electricity. In shaping the river through multipurpose dams during the New Deal and beyond, the federal government, Atomic Energy Commission, Bonneville Power Administration, aluminum companies, and private utilities seized control of the Columbia River. Not the least of the transformations was the production of vast amounts of plutonium and dangerous radioactive waste at the Hanford Production Facility in Richland, Washington. Throughout it all, Native Americans and settlers have been pushed aside for big business, and have also paid with their health.[2]

One of the more obvious stories of racism, xenophobia, and technology concerns the "iron chink" canning machine that accompanied the expansion of the salmon industry along northwest rivers. The inventor, Edmund Smith, built the device in 1898 to clean fish at canneries. Factory owners needed the machine to replace Chinese immigrant workers. In language about immigrants reminiscent of debates in twenty-first century America, Chinese workers were blamed for a recession that gripped the West Coast as cheap laborers who stole jobs from real Americans and refused to learn the language and customs. Congress passed the Chinese Exclusion Act (1882) to prevent further immigration. The Chinese faced violence in any event and were chased out of the Northwest. This left the canneries short of labor. Since speed was of the essence to process thousands of salmon during their annual run, salmon butchers were needed. But Chinese had left the Northwest rather than face violence, and those who remained often were old and could no longer work the hard, long hours required for butchering. Smith's fish-cleaning machine replaced the Chinese worker and came to be called the iron chink.[3]

While salmon struggled to swim past the thirteen major hydroelectric power stations built on the river, the potato found a home in irrigated agriculture and agribusinesses, and it floated to the center of the US food supply and nutrition system. The Bonneville Power Administration, a quasi-governmental bureaucracy based in Seattle, Washington, operates the dams. The BPA relishes and celebrates its electricity and water. It sees water and electricity as the key to a radiant future of productivity and jobs in aluminum smelting, plutonium production, and agriculture. BPA officials claimed in 1939 that the dams were the "power to make the American dream come true."[4] The dams, of course, changed the ecology of the Columbia River basin in a variety of ways—water temperature and chemistry, destruction of anadromous fish runs, ruination of American Indian patterns and lifestyle—but they enabled the development of the potato.

The Columbia Basin Project (or irrigation district) feeds 671,000 acres of agricultural land that were developed in the 1950s and 1960s.[5] By law, that land and inexpensive, federally subsidized water were designated for small farmers in parcels of 60 acres. But over time, through slight changes in the law and sleights of hand, the land was concentrated under control of big farmers and ultimately large corporate owners.

The potato spread from Maine, New York, and Pennsylvania to Idaho, Washington, and Colorado, along the rails of refrigerated wagons and with thirst quenched by growing industrial agricultural systems. Idaho and Washington account for more than half of the total US potato production. While planted acreage has declined from 3.9 million to 1.0 million acres in the twentieth century, agricultural research and development have led to increased yields. The yields came from new varieties known for uniform size and increased sugars needed for the French fry, and also from increased use of fertilizers and pesticides, integrated rail and truck refrigerated transport, and the nutrient-rich volcanic soils optimal for potato production, all on the foundation of irrigation. The number of farms dropped with concentrated production, equipment, and storage facilities, from 51,500 farms in 1974 to 15,000 in 2004.[6] While the 1862 Homestead Act specified 160 acres to a tenant, big businesses have acquired much more, on average in the potato region 3.5 times the acreage.

Some 300 commercial growers plant more than 160,000 acres annually, harvesting averages of 30 tons per acre, twice as much as the average yield elsewhere in the United States; Washington produces 20 percent of all US potatoes.[7] Much of the land is owned by ConAgra, One Hundred Circle Farms, McCain, and other massive companies. They have acquired substantial water rights. Intended to support farmers at subsidized rates and ultimately to benefit the consumer, instead prices have been negotiated for the long term that do not reflect market prices, and in spite of acreage limitations to limit water subsidies, pricing water below its value has created "considerable wealth for a certain class of recipient irrigators." The beneficiaries are those who owned the land at the time the project began to deliver water, while others who might wish to enter the market must pay market prices to receive entitlements to irrigation water.[8] In this way the consumer pays McCain and ConAgra even before the potato is harvested, and the system keeps new competitors out.

Such manufacturers as McCain dominate production—as a handful of companies dominate every sector of big agriculture: cattle, swine, chicken, turkey, corn, and potatoes. As McCain notes, "One of every three fries sold worldwide comes from McCain which translates into one hundred million servings of McCain French fries each and every day." That's a lot of starch, salt, and fat. McCain supplies homes, groceries, and restaurants with frozen potato and snack food products with "20,000 people and operating fifty-seven production facilities on six continents."[9] The French fry, in a "medium portion" of 380 grams (4.1 oz) at McDonald's, provides 29 percent of our daily fat, 12 percent of our saturated fat, 11 percent of our sodium, and some vitamins. You can burn that off if you jog for 43 minutes. And those Russet Burbank potatoes from which the fries come are sprayed with pesticides so toxic that farmers avoid the fields for five days after spraying, and the potatoes must be aired out to be processed.

We also have dehydrated potatoes, which have been sucked dry and then processed into powder, flakes, or granules, although they may also take the form of slices, cubes, or shreds. Along Fordist conveyor belts that resemble Model-T production, the potatoes are peeled, washed, cut, blanched or steamed, and dried in industrial dehydrators. They leave the industrial facilities in big trucks to travel to food services, restaurants, and cafeterias. They have the advantage of light weight for cheaper shipping, longer shelf life, and anonymity, and in the final insult to their organic nature, they ship in square boxes that suggest industry, not food. The dehydrated potatoes have lots of potassium and some vitamin B.[10] Add irrigation water, spices, and taxpayer subsidies, reconstitute, and you have American nutrition. Cattle and pigs enjoy eating the leftover glop.

State universities, potato councils that represent big growers, and fast-food joints are in it together to serve the potato's interest. University of Idaho researchers tout the potato's centrality to the state's future. They write, "Idaho is the nation's largest producer, packer, and processor of potatoes. . . . The state's growers produce about 30% of the US potato crop, but the Idaho potato industry is more than potato fields. Idaho frozen and dehydration processors produce 40% of US processed potato products, and Idaho fresh packers provide one-third of the nation's fresh potato shipments." They claim that potatoes provide 15,500 jobs and $3.4 billion in sales and when considering jobs outside of Idaho, add another 39,500 jobs.[11]

Potatoes are also crucial from a trade perspective: Frozen potato products comprise 60 percent of growing US potato exports. This means the world loves US potatoes, and there are fewer obstacles to market penetration for processed foods than fresh potatoes, although such issues as "high tariffs, irregular import regulations, maximum residue levels (MRLs), GMO and acrylamide concerns all have the potential to hamper frozen potato exports."[12] We are beholden to the potato.

The potato requires significant labor inputs to get to the table, too. The Secret Service and FBI joined with private police forces to harass farm labor organizers, as the US government long fought efforts of migrant farm workers to earn a decent wage that would allow them to provide schooling for their children, or protect them from inhumane conditions including long hours and pesticide exposure. The owners of big farms needed a huge supply of cheap labor, so that at times the businesses, which largely supported harassment, and the government worked at cross-purposes. In 1954 the government pursued the repugnant "Operation Wetback" to arrest and deport laborers who looked Mexican. Up to one million people were detained and often abused, many of them US citizens.[13] Eventually, through the Braceros program, illegal immigration, and businesses willing to look the other way, the burgeoning fruit and vegetable industry of the West Coast got its cheap laborers. But these men and women had few rights, except to supply consumers with cheap fruit and veggies. They toiled in poor conditions, and to this day suffer excessive and high exposure to pesticides.[14]

Aside from federal irrigation programs and cheap labor, potatoes required one more thing: good lobbying. Potato farmers have successfully pressed Congress to keep potatoes prominent in the American diet among children, even though these children need much less starch in their diets. However, the Women, Infant, and Children supplemental nutrition program has been unable to "free" the children from corporate potato lobbying. Starches are already well represented in diets of low-income people who, as all nutritionists, doctors, and parents know, need fresh fruit and vegetables, not potatoes or fruc. But in the absence of these other readily available foods, children have no choice. On top of this, Congress has sided with the rights of the potato to be ubiquitous, as if its needs trump those of children. Potato lobbyists have developed fine education tools to instruct children on the glories of the spud from an early age, for example, through instructional pamphlets provided to elementary school teachers at no cost. The materials give lots of advice on how potatoes are grown, how nutritious they are, and they provide suggestions about writing poems, songs, and jokes that highlight potatoes, but they offer no lessons on the history of irrigation, workers' rights and pesticides, or obesity.[15]

The result is the modern potato, the French fry, fatty diets, high blood pressure, skyrocketing cholesterol, expanding crowds of expanding people who have lost their wonderment about food. In a past life, last in the early 1970s, I ate fast-food French fries. Since that time I have been tempted to drop into fast-food stores and to ask for a discount since my tax dollars—and yours—subsidized the fry through those Bonneville Power Administration dams, Bureau of Reclamation irrigated fields, and Department of Agriculture research. Instead, I am struggling to find food without fruc.

The Origins of Fruc

Fruc is a self-augmenting product of industry as much as a food. It is an additive more than a nutrient. It is monopolistic capitalism, not public health. It grew out of the Industrial Revolution in the nineteenth century in an effort to establish uniformity and economies of scale, and to push the markets from local regions to the nation and then globally for agricultural produce. It is planting, harvesting, and processing. It is a railroad transportation network. Consolidation and overproduction have been the rule as food corporations have acquired many layers of the production process and successfully built monopolies with rails, especially in the upper Mississippi River basin from Ohio to the Dakotas.[16]

When tractors replaced horses and mules, rapid increases in the acres of land that could be used for crops other than oats for horses followed. In addition, tractors could easily tame contours of the land for any crop. By the 1950s widespread use of hybrid corn and other animal feeds led to the practice of CAFOs, or confined animal feeding operations, that have used antibiotics and drug fattening stimulation extensively. In 2013, 90 million cattle in the United States consumed nearly 60 percent of the nation's grain (95 percent of which is corn). The number of feed grain farms (those that produce corn, sorghum, barley, and/or oats) in the United States has declined in recent years, while the acreage per corn farm has risen, and the number of large corn farms (with more than 500 acres) has increased over time, while the number of small corn farms (with fewer than 500 acres) has fallen.[17] Advocates tout these economies of scale and seeming efficiencies as green revolution improvements. But the hybridization (and now engineering) of crops and fossil fuel–based pesticides and fertilizers has led to higher prices and increasing environmental costs. A pound of beef takes three-quarters of a gallon of oil to produce; a 1,250-pound steer takes nearly 300 gallons of oil. Seen in this light, fruc, like beef, is wasteful from the point of view of resource use, dangerous from the point of view of public health, and may be immoral.

Corn, like the potato, has become an industrial artifact, the result of hybridization and engineering, of experimental plots and federal and state research, of price supports and direct grants, and of such subsidized industrial goods as fruc and ethanol biofuel, the latter part of a misguided effort to turn corn into fuel independence when corn is environmentally unsound as a source of ethanol alcohol.[18] Corn benefits from the requirement—in the name of energy security—that each gallon of gas produced in the United States includes 10 percent ethanol; this pushed one-third of corn production into ethanol at a cost of $5 billion in 2007, on top of which production is protected by tariffs on Brazilian ethanol produced by cane more cheaply and with less environmental devastation. Indeed, corn subsidies in the United States totaled $84.4 billion from 1995 to 2012.[19] More than anything, corn is equal to billions of dollars in taxpayer money.

Not only have big businesses captured the irrigation water and land to foist starch upon us, but they have captured Congress whose antigovernment sentiments do not preclude massive payments to big farmers to bring us more sweeteners. The major beneficiaries of US agricultural farmers are commercial farmers, not family farmers, with an average income of nearly $250,000, and the only restriction is on couples who earn more than $1.5 million per year. The support programs date to the New Deal to protect the small family farmer, but Congress does not like the New Deal, judging by legislation of the last twenty years. It likes big business. Because of government policies and largesse, corn prices have increased many times over in the last decade and consumers pay the price in bread, milk, beef, chicken, and other food products.[20]

Corn is an Ellulian imperative. The process to produce HFCS that dates to 1957 hardly resembles what you would do with a food: take corn, then grind, obliterate, isomerize, ionize, and inject everywhere. First, the manufacturers obliterate corn to produce cornstarch, then convert the starch into a glucose syrup. This can be mixed with red food coloring "to make cheap and very convincing fake blood." They isomerize some of the glucose into fructose to produce HFCS-42 (42% fructose), HFCS-55 (55% fructose), and HFCS-90 (90% fructose). They produce HFCS-90 by passing it through an ion exchange column designed to retain more of its fructose component. HFCS-55, introduced in the late 1970s, is the most commonly used sweetener in US sugary soda waters.

Not only in Jamaica, but in the United States, sugar has had significant political, social, and environmental impact, much of it from agriculture, and especially from Big Sugar—a technology of growing, harvesting, processing, and refining that requires millions and millions of gallons of water annually. As noted in chapter 3, sugar is directly connected with slavery through the growing taste for the stuff in England and Europe. In the nineteenth century, sugar was England's leading import with per capita consumption rising from 4.5 pounds to 20 pounds in 100 years. Sugar is a gateway drug. It can be combined with chocolate and marijuana in brownies; it was combined with cocaine in Coca-Cola when the latter was first introduced in 1885.

And, why not produce it in a worthless swamp? From the early years of settlers, white folk have wanted to drain the Florida Everglades, a swamp at 4,000 square miles, by pulling the plug, or perhaps by digging canals, forcing water to the ocean, and reclaiming the land for sugar and other crops. As has frequently happened, government involvement both accelerated the "taming" of the land and enabled large landowners eventually to control it. And, as often happens, environmental degradation that will cost billions of dollars to remediate also occurred. A series of federal and state laws dating to the nineteenth century led to efforts to drain and settle the swamp. They were marked by corruption, financial failure, and no improvement in flood control.[21]

Like potatoes, sugar, corn sugar, and other products are Ellulian self-augmenting technologies. In the 1920s the US Army Corps of Engineers gained congressional support for increased budgets and massive projects, many in the name of flood control, including in the Everglades. These projects expanded during the New Deal. Later flood control projects resulted in the construction of 1,700 miles of canals, levees, and pumping stations. This enabled homeowners and sugar growers to move in. The sugar companies relied on migrant labor, immigrants, and the poor into the twenty-first century for the backbreaking labor of harvesting sugarcane with machetes; these laborers continue to be poorly treated.[22] The environmental impact—destruction of ecosystems and extinction of many species—looms large to this day, and Big Sugar, totally ignoring its profligate water, land, and pesticide use, insists that other folk historically are to blame for that damage, not responsible sugar.[23]

Yet somehow, in 1994, the Florida legislature passed the Everglades Forever Act to restore the swamp over several decades. Sugar agreed to pay only $320 million of the estimated $16 billion in restoration costs.[24] Of course, the plan foundered, and then in 2008 Governor Charlie Christ "announced a $1.75 billion deal to buy the U.S. Sugar Corp., including 187,000 acres (75,677 hectares) of farmland that once sat in the northern Everglades" that would give the state control of "nearly half the 400,000 acres (161,876 hectares) of sugar fields in the Everglades Agricultural Area (EAA) below Lake Okeechobee, although sources said U.S. Sugar would lease back its land for six years."[25] But Governor Rick Scott, who believes that anti-pollution laws hurt business, approved of plans to allow Big Sugar to pump pollution into the original Everglades in the name of "restoration."[26] These sweet twists and turns are not surprising. As the National Resources Defense Council has documented, the sugar industry "benefited from massive drainage and flood-control projects paid for by the federal government, cheap water prices, a federal program encouraging cheap labor from Caribbean workers, and federal quotas on sugar imports." Some would argue it bought these projects and influence. In the last decade alone the industry "contributed almost $3 million to congressional races."[27]

The Growth of the Sweeteners Industry and Lobby

In 1966, sweeteners added to foods in the United States totaled 113 pounds annually per capita. President Richard Nixon's agriculture secretary Earl Butz did a lot to see it grow to 147 pounds. In response to dropping farm income, Butz urged the planting of more corn and soy. The government subsidized the corn, and the rest is sweet history: urban residents have yet to be able to cut subsidies to the farm industry. Another reason for the rapid adoption of HFCS was the rising price of sugar at the same time as farm subsidies encouraged farmers to produce far more corn than Americans consumed. The result of excess supply and falling prices was the search for another market. Working together with such major multinational corporations as Coca-Cola, the corn industry pushed HFCS into beverages by the mid- to late 1980s where, according to *Obesity*, they average 59 percent of the sugar sweeteners in the nutritiously abysmal drinks. *Obesity* itself has grown since first published in 1993 to more than twenty volumes of articles, many of them about fruc.

In 1997, the corn sweetener industry used about 8 percent of the corn crop (20 million tons) and the fuel ethanol industry used about 13 million tons. But corn is more than sweeteners (dextrose, glucose, HFCS). It is starches (modified for food and industrial applications) and unmodified (complex carbohydrate for finishing, binding, and adhesive qualities), by-products (corn oil, germ meal, gluten feed for dairy and beef cattle, gluten meal which is a high-protein by-product for poultry, pet food, etc.), and other (dextrin, a roasted form of starch used as an adhesive and thickener, maltodextrin for bulking benefits, and steepwater from the water used to clean and soak corn in refining, which is used in feed and fermentation).[28]

Of course, fruc has many defenders owing to its safety in laboratory experiments and low cost and as a symbol of American know-how and can-do. American Chemical Society writer Mark Lesney hailed chemically arranged agricultural compounds as "the foundation of civilization." He added that chemicals "are intrinsic to the growth and health of today's highly managed agricultural systems (fertilizers and pesticides)." Let us not forget chemicals as additives, flavors, and for industrial processes (chemical feedstocks, pulp and paper).[29] In this light, the fructose and beef and feed—and potato—companies are chemical companies.

Chemists have undeniably increased crop output and dealt with pests. They have added arsenicals, copper compounds, chlorine, minerals, metals, and toxic oils. With the assistance of the US military, they first used airplanes for crop spraying in 1918. DuPont began insecticide research in 1928, and followed the discovery of DDT in 1938 with chlordane (1945), aldrin and dieldrin (1948), and heptachlor (1949). Almost all of the leading chemical companies have been involved in agricultural chemicals, especially pesticides: BASF, Dow, DuPont, Eastman, FMC, and Monsanto.[30] They produce purified and semi-purified agricultural components; such additives and preservatives as propionic, sorbic, and benzoic acids; calcium propionate, potassium sorbate; antioxidants; gelling, thickening, and stabilizing agents; and food ingredients of which fruc may be one of the most ubiquitous.

Like the potato, a handful of companies dominate corn and fruc. Five companies, Archer Daniels Midland Company, Tate & Lyle, Cargill, Corn Products Company International (CPC), and American Maize, sweeten our bounty, although they play hardball with price-fixing: ADM had 32 percent market share in 1994 followed by Tate & Lyle with 23 percent, Cargill with 19 percent, and CPC and American Maize at 9 percent each. ADM dates to the mid-1800s, when John W. Daniels and George P. Archer organized a linseed crushing business. They incorporated the Archer Daniels Linseed Company in 1905. In the 1970s, the company turned to corn processing in a Cedar Rapids, Iowa, plant; by 1972, their grind was 8,000 bushels per day. In the 1970s, ADM started making high fructose corn syrup and fuel ethanol. The company expanded rapidly in the 1990s with crystalline fructose, sorbitol, and maltodextrins, grinding 1.5 million bushels of corn per day. ADM produces roughly one-fourth of the more than 30 billion pounds of corn sweetener produced in the United States each year.

With the sale of Minnesota Corn Processors to ADM, concern grew among antitrust experts over market concentration in two key agricultural markets: ethanol and fructose. With MCP, ADM has a 40 percent share of the ethanol market. State subsidies for ethanol to MCP over the thirteen years through 2000 totaled $33 billion. And ADM and MCP are the two largest corn wet millers in the United States. ADM has $23 billion in annual sales, has 268 plants, and is the world's largest processor of corn, soybeans, wheat, and cocoa. Yet the US Justice Department permitted the merger with a stipulation that five independent competitors remain in the sweetener market.[31] ADM has already lost a lengthy price-fixing case (lysine, an amino acid used to enhance livestock growth, which ended with a $100 million plea-bargained fine[32]) and agreed to another settlement with soda pop manufacturers of $400 million in a price-fixing case involving HFCS.[33]

Cargill, Incorporated, the second largest supplier of corn sweeteners, began milling corn in 1967 in Cedar Rapids, Iowa, and over the next decades added several other milling plants in the United States, as well as plants in England, the Netherlands, Turkey, Brazil, Poland, and Russia. Cargill began production of fruc in the late 1970s, and also produces lactic acid, lysine, erythritol, and polylactide polymers. Cargill's Feed Division (Nutrena Mills, Inc.) merged with Royal Federated and Milling of Memphis in 1951. In 1952 they launched Carpolis, an integrated towboat and barge company for plying the Mississippi. By 1958 Cargill had expanded international operations to Switzerland and Antwerp (trade representation), sales to Argentina and soon to Peru, and later European and Brazilian markets. In the 1960s Cargill entered the chicken broiler and corn wet milling business, the latter in mills at Cedar Rapids. In the 1980s, as a multinational conglomerate, it sought new markets and new divisions: phosphate fertilizers, pork processing, nitrogen fertilizers, and petroleum products.[34]

Established in 1906, Corn Products International, Incorporated, in Argo, Illinois, has long maintained operations in Canada, Mexico, and South America. CPC also began producing fruc in the 1970s, and expanded to forty-two plants in twenty-two countries with subsidiaries, joint ventures, and alliances. In 2002, Corn Products International was the third largest supplier of corn sweeteners. The A. E. Staley Manufacturing Company, founded in 1898, has produced cornstarch since its beginning, and in the twenty-first century is one of the largest corn refiners in the United States, with capacity exceeding 600,000 bushels per day. Headquartered in Decatur, Illinois, the company product line includes sweeteners, starches, ethanol, animal feeds, and citric acid. It produces high fructose corn syrup, corn syrup, dextrose, and crystalline fructose for the food and beverage industries. These are used in baked goods, confectionery, fruit and vegetable processing, dairy products, and as a substrate for fermentation. (And let's not forget the billions of bottles and cans used for these products and their trade associations.)

Fruc is trade associations, from the National Corn Growers Association, to various state associations, and to those of the processed food industry, all of which have become quite adept at manipulating the media, the market, and the regulatory agencies, and which have massive budgets. The trade associations work at glorifying fruc among impressionable children the same way that cigarette manufacturers attempt to make smoking seem sexy and to addict children with an image of maturity and sophistication and the way the potato growers provide lesson plans to elementary school teachers. The Corn Refiners Association consists of virtually 100 percent of the market producers: Archer Daniels Midland Company, Cargill, Ingredion, Penford Products Company, Roquette America, Incorporated, and Tate & Lyle Americas. They glorify their industry as all-American, as the epitome of ingenuity with a grand history that is here to stay:

> The corn wet milling industry has been an integral component of American manufacturing for more than 150 years. As the industry grew from one small mill to a thriving group of corn refining companies, industry leaders formed the Associated Manufacturers of Products from Corn in 1913. In the 1930s, the group was renamed the Corn Industries Research Foundation recognizing the major contribution being made by the group to development of starch chemistry and technology. In 1966, the group was renamed the Corn Refiners Association reflecting the broad diversity of products produced by the industry.[35]

The associations advance the same arguments that all manufacturers of all adulterated food products do: that fruc is an "all natural, nutritive, versatile sweetener offering many benefits. It is very similar to sucrose (table sugar) and honey in composition, sweetness, calories and metabolism." They continue, it "provides energy, sweetness and moisture, and it enhances flavor and stability. It is found in numerous consumer foods and beverages due to its valued physical and functional attributes, including bran cereals, yogurts, dairy beverages, sauces, canned fruits, baked goods and condiments."[36] Between 2008 and 2013 the CRA spent $30 million on public relations. As food critic Marion Nestle writes, "Of that, $10 million funded research by James Rippe to prove HFCS is no different from sucrose (something you would learn from any basic biochemistry textbook). Mr. Rippe got a $41,000 *monthly* retainer from the CRA."[37]

State growers in Illinois, Iowa, Minnesota—all of the corn states—serve the same promotional activities. The Iowa Corn Promotion Board defends markets, funds research, and provides education about corn and corn products to ensure long-term profitability. Such boards have "check off programs" of $0.01 per bushel to accumulate bushels of money to lobby on behalf of members. Their goals include increasing the use of HFCS, working extensively with bottlers and soft drink companies, and educating consumers and food professionals.[38] They successfully increased the number of bushels going into sweeteners by hundreds of millions, and they have successfully tied federal support to corn through ethanol.[39]

But the associations are unwilling to permit more balanced and nutritious meals to be standard in schools if that means lower sales. Instead they proselytize sweetness among children. The Kentucky Corn Growers Association proclaimed, "Corn Is All Around Us!" in a brochure for fourth and fifth graders. The brochure indicated the importance of corn for the civilization of the continent (among Native Americans) and for the nation that pushed them aside. The brochure celebrated the pervasiveness of corn throughout the world except Antarctica, and its multiple uses, and especially its extensive applications in industrial foods like cereals, bread, juice, salad dressings, candies, jams, jellies, but also 3,500 different industrial nonfood uses/products: toothpaste, detergents, rubber tires, batteries, points and dies, body lotions, glues, lubricants. All these products were sweeter, stronger, more slippery, cleaner, tastier, healthier, friendlier, and softer.[40] And many of them led to more obese children.

In 2002, Americans ingested roughly 9.3 million tons of refined sugar and 12 million tons of corn sweeteners. However, between 2002 and 2008 the use of corn sweeteners in soft drinks, cereals, and a range of other products dropped 11 percent. In 2010, Americans used 8 million tons of corn sweeteners and 10 million tons of sugar. And a number of companies have stopped using corn syrup in some or all products, including Hunt's ketchup, Sara Lee, Snapple, Gatorade, and Starbucks' baked goods. But the downward trend is limited and faces the opposition of major lobbyists who will not retreat from their claims of safety and efficacy.

The Semantics of Calories and Nutrition

Trade associations play a major role in the debates over health and safety of various goods, services, and products. These associations defend the interests of the producers and ensure that accurate information about the products is widely available. They lobby the government for regulations beneficial to them in terms of marketing and sales. But they are no less a component of the technological system itself. Whether the National Potato Council or the Corn Refiners Association, they contribute to the sanctity of industrial food. In the words of Herbert Marcuse, they are, in this way, instruments of social repression. In *One-Dimensional Man* (1964), Marcuse described how modern capitalist and communist societies had created false consumer needs in a system of mass production where media and advertising encourage consumption. This led to a one-dimensional society where critical thought disappeared and powerful people shaped our desires. The creation of a lexicon of well-framed if misleading descriptions of various technologies is the result.

Because of the confusion—and bad press—that HFCS generates, the Corn Refiners Association has proposed renaming their substance "corn sugar." But whether this name is more accurate, it remains a sugar that consumers overuse and should not be using. Hence, according to some observers, we should be concerned about changing the name because of "corporate renaming efforts" that are "clearly deployed as strategies to confuse the public and inoculate industries in the wake of advocate attacks." The toxic sludge industry called the material that gets sprayed on thousands of acres of farmland every year "biosolids." BP spent perhaps $200 million to erase "Petroleum" from its name and added the green-spirited helios as its logo—although now many people rightly connect BP with the Deep Water Horizon Gulf of Mexico 87-day oil spill of 4.9 million barrels. Anna Lappe writes, "The Corn Refiners Association name change is another attempt to present high fructose corn syrup as natural—similar to and not any more harmful than table sugar."[41]

Yet how natural is it? Converting the glucose in corn syrup into fructose, as noted, is an industrial process and involves the use of genetically engineered enzymes. Fruc increases the shelf life of foods, which encourages its use in many processed foods, it's relatively cheap to produce, and is the epitome of economies of scale. Name changes reflect an attempt to put health disputes over fruc underground or into a dictionary and to obscure concerns about metabolism, heart disease, and added sugars. Marcuse would suggest that it is an attempt to sugarcoat consumer desires.

Americans have fought a long battle for food safety. That battle continues for food that is high in nutrition—or at least clearly marked so that consumers can judge how nutritious it is. The battle commenced more than one hundred years ago and was triggered by a fictional work, Upton Sinclair's *The Jungle* (1906) about the poverty and exploitation of immigrant workers in Chicago, many of whom worked in meatpacking plants where filth, disease, and injury were rife.

As with many other technological innovations that find their way quickly to the market, uncertainties remain about the safety and efficacy of fruc. Battles over these uncertainties take place in a regulatory environment where several federal bureaucracies—the Department of Agriculture and the FDA—engage public concerns, food producers and their trade associations, parents and teachers, and the US Congress to ensure a safe food supply. In this battle, manufacturers and food processors assure us that the world is sweeter than before fruc. Growing evidence that obesity and related diseases are connected with fruc consumption challenge this conviction.

According to a scientist who works for the sugar beverage industry, there is nothing wrong with fruc. Since the sugars are similar, many people assumed it would be metabolized in a similar manner. Most health experts believed that sucrose, fructose, glucose, and HFCS do not pose a significant health risk, with the single exception of promoting dental caries.[42] Although there was considerable speculation in the 1980s that fructose was responsible for several metabolic anomalies,[43] convincing proof that this was a significant health risk did not appear. To this day, many scientists vigorously oppose the notion that fruc poses any different risks than any other sugar. A 2012 meta-analysis of fructose studies published in *Annals of Internal Medicine* suggested that sugar calories in general, rather than specific sweeteners, were responsible for weight gain. But virtually all of the authors of that study had direct ties to fruc—with grants from Coca-Cola and other interested parties.[44]

Another specialist decided to downplay risk by pointing out that in one form or another, sugar has been a part of the human diet for centuries in fruit or honey; that it grows in equatorial regions from which it must be exported and where political and economic uncertainty has had an impact on its production. Sugar prices have fluctuated rapidly; hence processors saw fruc as a great alternative, as well as because it is a syrup, it can be pumped, it comes from dependable corn, and has kept relatively stable prices. Of course, these claims have nothing to do with public health. This author, John White, literally a one-man think tank, a member of industry-connected organizations that promote artificial sweeteners, deny global warming, ignore food risks, and rejected the 2004 paper of Bray and colleagues that linked fruc to obesity.[45] White condemned the Bray et al. paper for taking on a life of its own in the media. He criticized the hypothesis of a link by claiming that fruc and sucrose must be significantly different; that fruc must be uniquely obesity-promoting; must predict US and global obesity; and that a reduction of fruc would reduce obesity. The author notes, first of all, that fruc is only 42–55 percent fruc, that other researchers confuse fruc with common corn syrup and also with pure fructose. But he never addresses the question whether fruc and sucrose metabolize differently. He dismisses Bray's association of fruc with obesity for ignoring increases in total caloric intact or other added sweeteners. White dismisses the fact that fructose is rapidly taken up by the liver and bypasses a key regulatory step in glycolysis. He admits that "fructose malabsorption appears only to be a problem when too little accompanying glucose is present." He dismisses research that shows that fructose fed to experimental animals or human subjects in high concentration (up to 35 percent of calories) and in the absence of any dietary glucose produces metabolic anomalies by citing the 1994 Fructose Nutrition Review commissioned by the International Life Sciences Institute, a front organization for food and beverage, agriculture, chemical, and pharma companies to which he belongs, and by arguing that no one would eat a pure fructose diet. White minimizes the relationship between obesity and fruc by indicating that while obesity continues to worsen, the per capita intake of fruc calories has been in decline since 2002.[46]

Andrew Briscoe, III, president and CEO of the Sugar Association, and other sugary people, like the tobacco people before them, manufacture words and phrases to absolve sugar of any negative health indications and to put all blame for illness on the consumer. At the August 2005 American Sugar Alliance meeting in Sun Valley, Idaho, Briscoe blamed "obesity on people eating too many calories and not getting any exercise." He continued definitively, "Sugar is not a part of obesity issues."[47] Does he believe only fat, lazy people are the problem? Like other trade groups, the Sugar Association engaged in a duplicitous treatment of the "truth" about sugar and human diet. It funded so-called consumer groups, created false alarm among consumers, and generated false "scientific claims" about fruc. Briscoe knew that the data his group provided about fruc were wrong, yet the group publicly disseminated them. It gave the Citizens for Health more than $300,000 to fight fruc; Citizens actually is based at the DC law firm Swankin and Turner that houses several other supposed consumer groups.[48]

We know that there is a problem when technology trade associations get into a saccharine spat about who is sweeter. Sugar and fruc have entered such an expensive Marcusean fight. It dates to 2003, when Briscoe wrote the FDA to protest efforts of the manufacturers of fruc to claim their sweeteners were as natural as sugar and to rename their product "corn sugar." Briscoe's cookies were frosted because this meant more intense competition with the sugar industry, although Briscoe claimed it was a matter of "consumer deception."[49] Sugar sweetly sued fruc, although a federal judge in California judged it to be a "SLAPP" suit which means that "its primary purpose is to silence and harass the other side."[50]

Apparently responding to these developments, in June 2008, ADM, Cargill, and other HFCS manufactures, through their trade group, the Corn Refiners Association (CRA), launched a multimillion-dollar advertising campaign claiming that HFCS is "nutritionally the same as table sugar," "natural," and that "your body can't tell the difference." In their advertising campaign, which they ran without waiting for FDA approval, CRA began to call fruc "corn sugar."

More than 700,000 pages of documents connected to proceedings were filed in a federal lawsuit in 2011 in Los Angeles by the sugar industry against HFCP producers. Sugar alleged that fruc's campaign called "Sweet Surprise" was misleading when it claimed that "sugar is sugar" and that "your body cannot tell the difference between sugar and high fructose corn syrup." The documents indicate that both sides have paid researchers millions of dollars to defend their sweet teeth and attack the cavities in the other's arguments, making claims and counterclaims that one's product is natural and safe while the other is lying to hurt the consumer.[51]

Of course, corn refiners countersued Sugar in 2012 for its "spin and smear conspiracy" against fruc dating to 2003, and they enlisted the help of the Center for Consumer Freedom to protect fruc from slander. What is Consumer Freedom? A front organization for the corn manufacturers, although its website claims, "Founded in 1996, the Center for Consumer Freedom is a nonprofit organization devoted to promoting personal responsibility and protecting consumer choices. We believe that the consumer is King. And Queen." The Center notes that "a growing cabal of activists has meddled in Americans' lives in recent years. They include self-anointed 'food police,' health campaigners, trial lawyers, personal-finance do-gooders, animal-rights misanthropes, and meddling bureaucrats."[52] At least Consumer Freedom is protecting the rights of small children to eat sugar and benefit from the calories and glyceride spikes that may accompany sweeteners.

It was clear to investigative journalists—but not to the consumer—that the Sugar Association and the Corn Refiners Association were simply fighting for market share; both had directly blocked consumer education or misled the public, both had declared self-righteously a battle about free speech was at hand, CRA claimed that Sugar tried to silence them, while Sugar claimed that Corn was not a good sugar but connected with health issues.[53] As Eric Lipton of the *New York Times* reported, the conflict "demonstrates how Washington-based groups and academic experts frequently become extensions of corporate lobbying campaigns as rival industries use them to try to inflict damage on their competitors or defend their reputations against such assaults." The manufacturers fought over obesity, diabetes, terms like "natural," "sugar," and so on, and spent millions of dollars without dealing with the real issue: health and safety. For example, CRA spent about $10 million over a four-year period to help fund research being conducted by a Massachusetts-based cardiologist and health expert, Dr. James M. Rippe, whose research disputed any health consequences of fruc. Dr. Rippe denied that industry payments had any influence on his conclusions.[54] And he did not pay me to say that. The battle over names has little to do with public health and facts. It often has to do with who funded the research.[55]

Public Health and Fruc

One of the concerns of specialists in science and technology policy is how disputes among experts arise.[56] If "science" has to do with facts established through careful, replicable research in controlled experiments, then why do scientists sometimes disagree about the facts and about what those facts may mean? For example, we can understand how, during the Cold War, many scientists worried about the danger of the arms race for humankind and fought vigorously against it, while others in the West believed that the Soviet threat was significant enough to justify continued weapons development programs. Frequently, we must look attentively not only at the empirical aspects of a dispute, but at the political and economic aspects as well. The well-funded efforts of fossil fuel companies to create doubts about global warming indicate that some business leaders have shown a willingness to purchase research results with healthy doses of funding to specialists who share their concerns, even when scientific certainty has been established.[57] Remember also the efforts of the tobacco industry to mislead the public about the dangers of smoking.[58]

In similar ways, we must approach the protestations of the merchants of sweet teeth that their products have no adverse health effects with healthy skepticism. Americans face great pressures in the search for healthful food because of increased portion sizes, more fast food, less exercise, and fruc added to every product, while food manufacturers understand that such additives as salt, sugar, and flavorings captivate the consumer's taste buds.

On top of this, it is difficult for Americans to obtain complete information quickly—real, unadulterated facts, without additives—about the risks involved in consuming manufactured foods. Another ludicrosity—if they can make up foods, I can create words whose danger to your health is nonexistent—is the food pyramid. Founded on the monstrous notion that already obese people and soon-to-be-obese people need to consume from 2,000 (women) to 2,550 (men) calories a day, and that marbled fat manufacturers ought to help determine what we eat, the pyramid, a minor improvement from its predecessor the wheel, discourages public welfare. One of the reasons that food contents and food labels are confusing is that the US Department of Agriculture and the FDA have different responsibilities for safety—and different missions. In 1991, the FDA commissioner, David Kessler, determined to begin a battle against false health claims on food labels, for example, the misuse of terms like "fresh" and "no cholesterol." But his actions angered food trade groups for "confusing" the consumer—or in this case for providing the consumer with more information. The Department of Agriculture, as noted here, has the reputation of working closely with producers to ensure sales, not the least through direct subsidies and price supports. Hence, and with the instruction of Congress, it has framed commodity, school lunch, and other programs often to meet the demands of industry. And Congress has done its best to control the department's Food Safety and Inspection Service, to a great extent by cutting its budget for inspection to limit their scope and effectiveness and the numbers of employees. Its functions are fragmented, and many inspection programs rely on food processing personnel to serve the inspection role and make claims about compliance and safety. (In any industry, should the regulated be the regulator? In this case, the result is faster carcass processing, not full inspection.)[59] This is ultimately a case of agency capture, where Agriculture promotes and regulates a product.

But Kessler succeeded in pushing a revision of nutrition facts labels in 1992; previously they had been tiny and confusing. Kessler called this a "battle royale." He said, "Every change is a battle with the food industry." Yet he concluded, "The food label that we implemented—did it harm the food industry in any way? No. In fact, I'm sure they profited from it."[60] Kessler also attempted to go to battle with the American tobacco industry in the 1990s to regulate exposure to tobacco smoke.[61] The problem was that Congress, some of whose members accept checks from lobbyists on the House floor, including John Boehner who later served as Speaker of the House,[62] has fewer pockets for protecting the public. It therefore prevented the FDA from regulating tobacco until 2009, and has little stomach for regulating food. Kessler has been tireless in pursuing food safety, documenting how foods high in fat, salt, and sugar alter the brain's chemistry in ways that compel people to overeat. Kessler toured dumpsters and other "sources" to determine just how much of the stuff was in our foods.[63]

Ultimately, there may be no problem with fructose except in large quantities. Yet recent evidence suggests that fructose consumption is linked to weight gain and type 2 diabetes.[64] Another study indicated that countries that use large amounts of fruc in their food may be helping to fuel the global epidemic of type 2 diabetes. Researchers from the University of Oxford and the University of Southern California found a 20 percent higher proportion of the population have diabetes in countries with high use of the food sweetener compared to countries that do not use it.[65] Consuming fruc-sweetened beverages increases adiposity in mice, while modest fruc beverage intake can cause liver injury and fat accumulation in marginal copper-deficient rats.[66] I have a sweet tooth, but I am not a rodent.

The sweet guys at sugar and fruc will debate the facts and pay their consultants to contribute to further debate, while Americans will get fatter, children will get sick, and the trade associations will fight to retain their rights to sweeten foods unnecessarily over the rights of children to consume good food. Efforts to limit sugar intake, or generally to provide good foods to schoolchildren, have been buried under a pile of rhetoric and trade association contributions to congressmembers. The French fry, pizza, and sugar seem to have meant more to lawmakers than did public health. Congress, which since 2000 has been known solely for the inability to pass significant legislation because of partisan gridlock, found it possible in thirty days to reverse Department of Agriculture efforts to ensure a balanced diet in school lunches. In November 2011, Congress overturned the first commonsense public health change to diet in the $11 billion school lunch program in fifteen years; the change was intended to reduce childhood obesity by adding more fruits and green vegetables to lunch menus. The rules, proposed in January 2011, would have cut potatoes, cheese, fat, salt, and sugar in children's school lunches. But a House and Senate "compromise" on the agriculture spending bill blocked the department from using money to carry out any of the proposed rules—and thereby doomed children to face the risk of more obesity.[67] Even the School Nutrition Association, an organization that represents "55,000 school nutrition professionals working in cafeterias nationwide," sought a delay in implementing the legislation for the "Healthy, Hunger-Free Kids Act" for at least one year to give schools "reasonable flexibility" to bring healthier meals to children. Representing food interests, not the interests of schoolchildren, the Association wanted the delay because they are worried about profits.[68] If they are nutrition "professionals," then they ought to know better. Who represents the children?

The good news is tempered by the bad. Finally, even the USDA website now has information on obesity prevention.[69] Meanwhile, the Corn Refiners Association spent millions of dollars on a public relations campaign about the natural goodness of fruc—or nine times more than the Centers for Disease Control and Prevention allocated that year for its entire fruits and vegetables promotion program, while 17 percent of American children are obese.[70] Nearly half of all "vegetable servings" that Americans consume come from tomatoes, iceberg lettuce, and potatoes, the last mostly as French fries and potato chips.[71]

Get the Fruc Out of My Food

Many foods that reach the tables of modern consumers are products that epitomize the industrialized manufacture of taste, texture, and nutrition. They grew out of the economic and political power of interconnected technological systems from engineered rivers and irrigation systems to massive landowning corporations. Taking advantage of subsidized water and electricity rates, of congressional programs to improve shipping, of price supports and subsidies, the companies built large farms—agribusinesses—that together produce the lion's share of a number of commodities—chickens, beef, potatoes, corn, and so on. Many of these commodities are monocultures, selected and engineered to work with various chemical inputs to produce the perfect French fry or sweetest fruc. The trade associations that represent the interests of the commodity producers have, in many cases, captured the regulatory impulse and framed the debates over the "facts" of the health and safety of the US food supply. The result is engineered food that has captured our taste buds with salt, fat, and especially sugars.

Fruc is now everywhere. Look at the labels. A Subway-brand sandwich includes enriched wheat flour (with less fiber and vitamins—that must be added back in—and a higher glycemic index than wheat flour with added fruc) and meats, poultry, and cheese with dextrin, dextrose, corn gluten, modified food starch, maltodextrin, corn starch, and hydrolyzed corn protein, and condiments, all with sweeteners. If you need even more fruc, try the cookies.[72]

Fructose metabolism produces triglyceride spikes in blood and may also have a connection with diabetes. It contributes to obesity. And it is junk food, nutritionally vacuous, that was marketed to children directly in public schools through vending machines with the complicity of adults. It took until the summer of 2013 for the US Department of Agriculture to establish "Smart Snacks in School" nutrition standards so that any food sold in public schools must meet (and not exceed) calorie, fat, sugar, and sodium limits. While schools can still sell brownies and cupcakes at bake sales and sporting events, snacks sold during school hours cannot exceed 200 calories and must be either full of whole grains or primarily contain fruits, vegetables, dairy, or protein. (The new standards did not apply to foods brought to school in bagged lunches, or for activities such as birthday parties, holidays, and other celebrations.) The standards allow chips, but in such healthier versions as baked tortilla chips, reduced-fat corn chips, and baked potato chips, and granola bars, popcorn, fruit cups, and calorie-free flavored water instead of the usual list of corporate snack foods and soda.[73]

But mostly, calls for concern and regulation remain cries for help lost in tall stalks of corn. Fruc is most detrimental to the poor. They seem always in every way to bear the burden of life in the technological utopia we have created. How much do the sugar industry and the corn industry get in subsidies from the federal government? Got milk? What is the other white meat? Is it good that cooperatives stabilize prices and raise profits? How do commodity programs cost taxpayers billions of dollars which go directly to businesses? If the market is absent, don't consumers pay higher prices? What is "mechanically separated" meat or chicken? What is fruc? It's a large-scale technological system that should require a "Fructura" symbol be put on packages of processed food.

In the aftermath of Hurricane Katrina—a "natural disaster"—people walk through the New Orleans floodwaters to get to higher ground, August 31, 2005. Natural and anthropogenic disasters reveal the intimate interconnection among technology, nature, and the human decision making that puts individuals at great risk. Courtesy of Federal Emergency Management Administration. Photo by Marty Bahamonde.

5

Technology and (Natural) Disasters

You Cannot Fool (Mother) Nature

What has happened down here is the wind has changed
Clouds roll in from the north and it started to rain
Rained real hard and rained for a real long time
Six feet of water in the streets of Evangeline
—RANDY NEWMAN, "LOUISIANA 1927" (1975)

When an earthquake strikes or a monsoon hits, people call these events natural disasters. When flooding or mudslides inundate a small village, journalists claim that Mother Nature has unleashed her fury. They inevitably declare that nature is "mother," attributing a kind of capricious, female behavior. But natural disasters are hardly the stuff of nature behaving in an unpredictable fashion. Centuries upon centuries of local knowledge about the regular and predictable patterns of weather, decades upon decades of scientific data that reveal well-known trends by month and week, and now weather satellites, seismographs, and other technology that show in real time approaching storms and tsunami risk indicate a more complex picture of the terrible natural disasters that destroy entire communities. Yes, gender—and race and class—play a role in disasters, not because of the alleged capriciousness of female nature, or uncertainty about the extent of meteorological and geological risk, but because those of the lower classes, people of color, and women and children tend to suffer the consequences of disaster and slow pace of reconstruction much more than the politically connected and economically fortunate. Social structures, political decisions, building patterns, scientific understandings, and so on all contribute significantly to natural disasters in some way.

Many natural disasters appear to arise out of gradual or cumulative processes, for example, a wildfire that spreads rapidly only when sufficient combustible material has gathered on the forest floor over decades. In cases like this, because the disaster will occur after some threshold of natural conditions exists, planners, policy makers, scientists, engineers, and public groups believe that they may preclude or minimize that disaster event, for example, by engineering rivers to prevent floods, building breakwaters to protect harbors from coastal storms, or establishing fire suppression programs in forests. Many other disasters seem to be beyond human intervention: earthquakes, volcanoes, tornadoes, although through building codes, the stipulation of exclusion zones, and emergency warning systems they may minimize loss of human life and the destruction of property.

Admittedly, some disasters seem to be more "natural," a hurricane or monsoon, earthquake or volcano, for example, and others more "human," or as we say "anthropogenic," for example, an oil refinery fire, an oil tanker running aground, or a reactor explosion. Yet whether disasters arise from cumulative causes or understandable but unpredictable ones, both human and natural agency play a role, and we ought to avoid seeing disasters in binary fashion, but rather as something complex that involves humans and nature. To put it simply, without humans to apprehend nature, there are no natural disasters.

Indeed, natural disasters involve human presence—their villages, homes, cemeteries, businesses, factories, schools, and hospitals—and their science and engineering. The way they organize agriculture, build structures, and attempt to control nature to limit its capricious fury is central to natural disasters. The establishment of centrally funded large-scale research organizations in the mid-nineteenth century went hand in hand with centrally funded nature transformation projects for canals, roads, and railways, hydroelectric power stations, irrigation systems and so on, precisely because scientists, engineers, policy makers, and others believed it possible and prudent to control nature. To this day, science and engineering remain a crucial aspect of disaster history. To call something natural without seeing humans as part of it ignores not only the epistemological, but also the physical and psychological. It also avoids recognizing the interaction of politics—of class, race, and gender—with science and nature in all disasters.

How Natural Floods Are Human Disasters

Farmers have long recognized that the floodplains along rivers have the most fertile land, and they may expect—and welcome—spring floods for spreading nutrients across their fields. Yet for a variety of reasons, as societies have become increasingly urban, with cities frequently growing up on rivers because of the benefits of inexpensive transport and fresh water, annual floods have become a natural disaster. Water inundates factories, office buildings, and homes. Engineers turn to "flood control," and indeed have developed academic fields of specialization (hydrology, civil engineering, and others) to control floods. They erect dams and levees, they dredge channels. And, of course, floods recur. Seeing a river, civil engineers see a quiet plain, a place to build a roadway or a railroad. Yet during the wet season the roadway will flood and a hapless driver or two will drown. Cities like St. Petersburg, Russia, built on a swampy coast, will experience extensive and dangerous floods; entire countries like Bangladesh, in which people live near rivers, will be inundated.[1]

The first multi-lane, limited-access parkway in North America, the Bronx River Parkway, that parallels the north-south flowing Bronx River, was completed in 1925. Designers intended it to demonstrate that nature and the automobile went hand in glove, for they sought to use the parkway to permit middle-class Americans to drive an automobile through nature, preserving thousands of acres of abutting land as a park in Westchester County, New York, simultaneously hoping to promote commerce through efficient traffic movement. The parkway system built around New York itself has been an object of social engineering, with bridge underpasses designed not only to be aesthetically pleasing, but to prevent buses from using them, and hence, the thinking was, making parkways accessible to white people who had automobiles, not black people who likely were limited to public transport—including buses.[2]

Building along the floodplain hardly has been a blessing, damaging fragile ecosystems, and flooding out the road every spring, and often after a heavy rain, for there is no way to prevent it. Wetlands that used to soak up the excess water have been reclaimed or paved. In March 2007, one such flood left the police exasperated, while attributing blame to Mother Nature.

> According to the Westchester County Police nothing can be done to alleviate the [March 2007] flooding. They are simply waiting for Mother Nature to take its course and waiting for the water to recede. A spokesman for the County Safety Department claimed that inadequate storm drains are just part of the problem and Scarsdale's Deputy Village Manager Stephen Pappalardo agrees. The Bronx River is filled with silt and therefore has limited capacity to capture excess water. Dredging the river to remove the silt is an environmental concern as it would upset the ecological system and wildlife.[3]

But politicians will always suggest an expensive solution to save wealthy Westchester County residents and their expensive vehicles. Westchester County

Executive Andrew Spano proposed electronic gates that would control access to the parkway during floods, although he offered no construction schedule or cost estimate, and nothing has happened since—except more flooding.[4] As numerous skeptics have noted, Americans drive their cars on parkways, and they park their cars in driveways. Sometimes, they drive into deep water and simply lose their cars.

The engineering of the Mississippi presents an intriguing case of the complexity of the natural/anthropogenic conundrum, and raises the specter of racism in natural disasters. The Mississippi overflows its channel each spring, carrying vast amounts of silt into the floodplain and also downstream where it extends the delta. Since the mid-nineteenth century, and especially in the twentieth century, the Army Corps of Engineers, unencumbered by modesty of vision, has built canals, locks, and levees to improve transport, control the river, prevent flooding, and protect property. In concert with the Bureau of Reclamation and private citizens, it has transformed the abutting wetlands into farms and towns. But in the effort to prevent floods, they have turned 100-year events into 50-year occurrences, 50-year floods into 25-year occurrences, and so on, and they have encouraged building and the location of industry and farms into areas that will eventually flood—with damages rising into the billions of dollars. Channeling the river and building levees funnels more water moving with higher energy and greater scouring power downstream, so that when water spills over, it does so with greater volume and force; building of roads, bridge abutments, and other structures along reclaimed land that no longer acts like a sponge exacerbates the situation. In an extreme case, the Corps attempted to force the flow of the Mississippi River into a tributary, the Atchafalaya, for flood control to protect New Orleans.[5] The engineers claimed they had protected New Orleans from even class 4 or 5 hurricanes from the Gulf of Mexico and from spring floods coming downstream. As we know from Hurricane Katrina which struck with class 3 force, these were faint dreams. Perhaps we should build only three things in floodplains: golf courses, cemeteries, and shopping malls. Golf courses will dry out and may be swept clean. Cemeteries are underground, and their occupants do not vote. And the loss of millions of dollars of needless consumer goods should be a source of joy, not of sorrow, except for insurance companies.

Both the epic floods of 1927 and Hurricane Katrina indicate that these "natural disasters" play favorites: those who suffer the greatest losses are people of color and people from the lower classes. The horrible Mississippi floods of 1927 led to another round of legislation and budgetary appropriations to seek control of the Mississippi. But this natural disaster should be known more for the way in which race played out in human disaster. At the insistence of wealthy property owners, African Americans were forced at gunpoint by the National Guard to stay on levees at Mounds Landing and continue filling sandbags. As many as 200 of these workers drowned when the levee finally gave way. In Greenville, Mississippi, in the heart of the Delta, more than 10,000 people took refuge on a levee protecting the town, one of the few areas of dry ground. Their condition grew desperate. Plans were made to evacuate all of Greenville's residents. However, local landowners protested, complaining that sending away all the area's blacks would deprive them of their primary source of cheap labor. At the last minute, plans to evacuate African Americans via steamers were cancelled. All the blacks in the area were forced to leave the levee, one of the few areas above water, where they were made to load and unload supplies without pay. Those who refused to work were denied Red Cross relief rations, and others were beaten with impunity and held as virtual prisoners. Similar conditions prevailed in other camps.

Class and wealth played out in the floods in untoward ways as well. Some 10,000 people lived in St. Bernard Parish north of New Orleans. New Orleans businessmen agitated to destroy the levees there to dissipate the flood before it reached the city. They secured the approval of the administration of Republican President Calvin Coolidge, with the proviso that they compensate people for property losses. But the Reparations Commission set up to administer claims was run by New Orleans bankers and businessmen, and most people received nothing for their losses. The National Guard carried out the forced evacuation of St. Bernard Parish, and those residents who had nowhere to go were housed in a large warehouse in downtown New Orleans, whites on the fifth floor and blacks on the sixth floor.[6]

Hurricane Katrina's great human, environmental, and property costs—more than 1,800 people died and losses topped $81 billion—indicate the power of the storm, the engineering considerations that contributed to the losses, and the cynical political considerations that exacerbated the situation—as they had in 1927. First of all, New Orleans flooded when the levee system protecting it failed catastrophically, in many cases hours after the storm had moved inland. Studies by Louisiana State University and the US Army Corps of Engineers confirm that steel reinforcements on failed levees only went half as deep as they were supposed to go. Not only were the levees not built to withstand a category 4 hurricane, despite repeated warnings that such a storm was inevitable, but due to shoddy construction they could not even withstand a hurricane of lesser force. The construction flaws virtually guaranteed that they would give way in the face of a storm with the power of Katrina—a slow-moving category 3 storm when it hit land. Ultimately, 80 percent of the city and surrounding parishes flooded. The floodwaters remained for weeks. Pumping out New Orleans took forty-three days, with sewage, bacteria, heavy metals, pesticides, and oil transferred into Lake Pontchartrain. Mississippi beachfronts suffered greater property damage.[7]

Accused of insensitivity, incompetence, and ultimately racism, the Bush administration responded tardily and with inadequate supplies. Michael Brown, chief of the Federal Emergency Management Administration, demonstrated incompetence, although President George Bush congratulated him for doing a great job ("a heck of a job, Brownie"). He did not ask the president to include the coastal parishes of Orleans, Jefferson, and Plaquemines in a disaster decree, and he ignored Louisiana Governor Kathleen Blanco's letter asking him to do so; she had requested assistance for "all the southeastern parishes including the City of New Orleans" and fourteen parishes including Jefferson, Orleans, and Plaquemines. Granted, Mayor Ray Nagin of New Orleans and Governor Blanco faced criticism for failing to order people to evacuate until late in the disaster when traffic jams and the approaching storm made evacuation nearly impossible. But there was no provision to help the elderly black residents of the hardest hit regions to get out, or to assist them in the cleanup.[8] And President Bush, apparently not knowing disaster history, nor knowing Randy Newman's song, says he did not know the hurricane would hit so hard, and how could he have known? Perhaps he did not know, but there was no excuse for the miserable response of federal disaster authorities.

Katrina had a profound impact on the environment, both because of the rich biodiversity and number of ecosystems in the region and because of the presence of a large number of petrochemical and other facilities in the Mississippi delta. Storm surge caused substantial beach erosion, obliterated barrier islands, and ruined habitat and breeding areas for marine mammals, such birds as pelicans and ducks, turtles, and fish. The damage from Katrina forced the closure of sixteen national wildlife refuges. The hurricane triggered forty-four oil spills with a total of 7 million US gallons of oil, with four leaks of roughly 1 million gallons or more. (In 2011, opponents of President Obama briefly criticized his administration for doing *too much* in preparation for Hurricane Irene, since the storm never lived up to meteorological predictions. Given the contrast with the Bush administration's incompetence responding to Katrina, this kind of commentary had no resonance among thinking people. Brownie himself had the audacity to criticize President Obama for moving *too* quickly on Hurricane Sandy.[9])

Policy makers and property owners themselves are partly responsible for the property losses in natural disasters. The US government encourages rebuilding in areas prone to be hit repeatedly by storms. Although it has cut budgets and tightened requirements on which projects can be funded, the US government, through FEMA and the Corps of Engineers, will support municipal and state requests for sand replacement.[10] Of course, sand replacement projects fly in the face of knowledge that beaches are dynamic ecosystems and that "replacement" almost always accelerates ongoing processes of beach decline due to wind, rain, storms, and erosion. The process must be repeated once it has begun. But state governments make the case that tourism will decline if beaches disappear, and so they count on the federal funding.[11]

Politicians who decry federal funds are happy to get millions of them for their beaches. Governor Chris Christie (R-NJ) proudly proclaimed that he had "restored [state] beach replenishment funding to its full level." The controversial practice "sucks sand out of the sea, or carts it from other locations, and dumps it on the shore. In places like New Jersey, the natural pattern of the wind and waves would, without the artificial efforts of engineers, whittle away at beaches and threaten the houses that sit along them." Yet beach replenishment often must be repeated every few years and encourages private investment where there should be none. An informed opponent of this unsound practice said, "In fact, in New Jersey, significant expansion of coastal 'McMansions' follows beach fill projects regularly." He continued, "The public pays to create a beach, and property owners and local politicians rush in to build huge waterfront houses where small beach bungalows once stood."[12] Politicians who welcome federal largesse also praise sand replacement: Governor Andrew Cuomo (D-NY) announced a $207 million plan "to dredge millions of tons of sand off the south shore of Long Island and spread it along the beaches and dunes," with the Army Corps of Engineers promising this will "stabilize Fire Island and reduce the storm surge hazard for the mainland." But, as the science shows, it will do neither.[13]

Insurance on one's property and belongings also shapes natural disasters. For many years in the United States, for example, federal programs for flood and disaster insurance enabled individuals—usually individuals of good means—to build—and to rebuild—in disaster-prone areas, those subject to hurricanes, fires, flooding, and so on. The owners and their insurance companies knew that subsidies made it possible to build on top of the ocean in coastal Carolina, Cape Cod, in river valleys, in the canyons around Los Angeles. Of course building in these areas was both subject to and contributed to more erosion, flooding, and so on.

Many of the threats that residents of coastal regions face come from global warming centered on significant—and accelerating—sea level rise and more intense storms, both of which threaten their homes and livelihoods.[14] The resulting destruction has been a common occurrence in other cities of the world, for example, Dhaka, Bangladesh, where intense monsoon storms hit frequently. But Americans finally understood the danger when Hurricane Sandy hit the New York City metropolitan area in October 2012, causing at least 268 deaths and total damage of $68 billion.

Earthquakes and Engineers

One of the most terrifying natural calamities is the earthquake. Tens of thousands and even hundreds of thousands of people have perished in single quakes and the aftershocks. The usual culprit, whether in Spitak, Armenia, in 1988, Bam, Iran, in 2003, or Sichuan, China, in 2008 where 70,000 people died, is the collapse of domiciles, places of work, schools, and hospitals on inhabitants. The nature of earthquake danger grew more complex in the twentieth century. Mass-produced housing and office buildings, often high rises, while meeting urgent demands for inexpensive structures and short time frame for construction, have set the stage—unintentionally—for disaster. In some cases, engineers and construction companies have deliberately cut corners. In Bam, nearly 27,000 deaths occurred, a number that was exacerbated "by the use of mud brick as the standard construction medium; many of the area's structures did not comply with earthquake regulations set in 1989."[15] In the Sichuan case, government officials estimated that more than 7,000 poorly built schoolrooms collapsed. This would never happen in elite communities or in China's gleaming modern cities. But why were they even built in earthquake-prone regions? Apparently, builders cut corners by replacing steel rods with thin iron wires in reinforced concrete, used inferior grade cement, if any at all, and skimped on bricks. Local residents, without the power and access to resources of their urban compatriots in Beijing, took to calling these structures "tofu-dregs schoolhouses." To make matters worse, because of China's one-child policy, many families lost their only child in the disaster.[16]

Similarly, the Spitak earthquake was a direct result of the efforts of Soviet officials in the Brezhnev era to hold back on expenses for social services, housing, and so on, while focusing investment on heavy industry and the military. The December 7, 1988, earthquake killed at least 25,000 people, many of whom were crushed in buildings that collapsed like decks of cards. Apartments, schools, and hospitals were particularly vulnerable. Without fully functioning hospitals, many injured died because of exposure to the winter cold and snow. Subsequent study indicates that the destruction was mainly to reinforced concrete buildings, often made out of forms that were fashioned simply, or buildings of mixed construction that combined masonry with reinforced concrete. Those buildings dating to the nineteenth and early twentieth centuries using mostly masonry with timber floors and roofs suffered only slight damage, and they had survived a 1936 earthquake as well. One expert reached the inescapable conclusion that traditional construction methods and materials had stood the test of time, while mass production techniques contributed to the extent of the earthquake disaster.[17]

Anthropogenic Disasters: Engineering Hubris

Many disasters appear to be entirely anthropogenic. They require human agency in some direct way: the poor design of a technology, the confluence of unexpected events, technological failure, an on-the-job mistake. (The legendary so-called Flaming Rat Case [*United Novelty Co. v. Daniels*, 42 So. 2nd 395, Mississippi, 1949] was such an on-the-job mistake, where a worker had regularly used gasoline as a cleaning agent in a room with a gas-powered heater with pilot light, and apparently, although not proven convincingly, a rat got doused, and moved across the room to the heater. The employee, and apparently the rat, died in the explosion.[18]) Yet even for many of these disasters—or disasters waiting to happen—human/nature interactions are paramount, and engineering hubris is often a root cause. Designs of technologies without safety redundancies, overconfidence that engineers understand the complex interactions between components under stress, siting of large-scale facilities near population centers or in areas known to experience extremes of climatic or geological conditions, for example coastal Florida and the Gulf states with hurricanes, the California coast and its active seismic faults and its potential vulnerability to tsunamis. Building pipelines in circumpolar regions and wells with offshore oil platforms begs for an accident with significant environmental impact and loss of human life. These disasters often acquire emblematic names: Chernobyl and Fukushima (discussed later) and Bhopal.

On the night of December 3, 1984, a Union Carbide India Limited pesticide plant in Bhopal leaked tons of toxic gases including methyl isocyanate that killed at least 4,000 if not four times as many people. As of 2006 the Indian government indicated hundreds of thousands of injuries including roughly 4,000 permanent disabilities. Injuries include neurological disorders, blindness, skin problems, and birth defects. Union Carbide has denied responsibility for the accident, claiming sabotage, while official reports indicate poor management and inadequate maintenance led to the leak and explosion. Union Carbide agreed to hundreds of millions of dollars in payments, but never admitted liability. In 2010, seven ex-employees, including the former UCIL chairman, were convicted in Bhopal of causing death by negligence and sentenced to two years' imprisonment and a fine of about $2,000 each. The chairman refused to return to India from the United States, and the United States would not extradite him; he died peacefully in a nursing home in Florida in September 2014. But it seems clear that inadequate safety systems, the lack of backups, and poor training all conspired in this disaster, as they nearly always do in chemical and petrochemical plant accidents.[19]

Frequently cost-cutting measures join hubris in designs, for example those for ocean-going oil and natural gas tankers. That is to say, technological momentum, human error, and nature have come together in another series of disasters: oil tanker spills. Rough weather may be a frequent cause, but so is the fact that supertankers (VLCCs, or very large crude carriers, and ULCCs, or ultra large crude carriers) have grown to 400 meters in length, may require more than 2,000 meters to stop, and must maneuver into harbors through narrow reaches to unload their cargo. Supertankers are disasters waiting to happen because of their crude carrying capacity. In the effort to avoid accidents, they are now highly automated with computers and GPS satellite navigation to preclude human error; a small crew operates the entire vessel and one person "drives." Yet, when loaded, the ship is mostly underwater like an iceberg. This is the deepest and slowest of big ships; currents can push them off-course more easily than they can push faster vessels.

After World War II, a typical tanker was 160 meters with a capacity of 16,500 deadweight tonnage (DWT). The ULCCs built in the 1970s had a capacity of

500,000 DWT—thirty times larger. Sonar, new steels, excess peacetime ship-building capacity, and growing demand for oil gave impetus to their construction. Another factor leading to their construction was the Suez Canal dispute, along with nationalization of Middle East oil refineries, that led oil companies to build ships to avoid political uncertainties that would threaten oil supplies.

Large tankers, built with single hulls, move oil cheaply, now some 2 billion barrels annually. Precisely the absence of double hulls and the difficulties in maneuverability led to the *Exxon Valdez* oil spill on March 24, 1989. The *Exxon Valdez* struck a reef in Alaska's Prince William Sound and spilled 260,000 to 750,000 barrels (41,000 to 119,000 m^3) of crude oil, the largest such spill in US waters until the 2010 Deepwater Horizon oil rig disaster (4.9 million barrels or 780,000 m^3). The oil covered 2,100 km of coastline and 28,000 km^2 of ocean with long-term, significant, and irreversible damage.[20]

Only this terrible accident led to greater regulation of supertankers. In 1990 the US government passed the Oil Pollution Act. Subsequently regulations of the International Convention for the Prevention of Pollution from Ships effectively mandated double hulls for newly built oil tankers larger than 5,000 deadweight tons to prevent or reduce the possibility of an accident because of grounding or collision. Yet the 1990 US act limits liability to $10 million and fines are small compared to the billions of dollars the companies earn.[21] And there will continue to be human error and unsettled weather. Further, all tankers will have to be converted (or taken out of service) when they reach 30 years of age. Why? "This measure was adopted to be phased in over a number of years because shipyard capacity is limited and it would not be possible to convert all single hulled tankers to double hulls without causing immense disruption to world trade and industry." That means that for another decade single-hulled ships lurk out there,[22] as natural disasters waiting to happen.

Global warming resulted to a great extent because of fossil fuel burning and greenhouse gases that fill the atmosphere and trap the heat. The polar ice fields are melting. Yet the search for oil and gas goes unabated, and the United States has turned to more and more aggressive—and more and more dangerous—methods to secure it. These include invading Iraq to try to stabilize the Middle East and other multibillion-dollar foreign policy adventures centered on the US thirst for petroleum; drilling oil wells in such fragile environments as the Arctic and offshore; and more recently, fracking (hydraulic fracturing) with hazardous pollution reaching drinking water. Oil spills from the *Exxon*

Valdez tanker accident to tar sands and pipeline pollution in Alberta, Canada, to deepwater locations in the Gulf of Mexico and Alaska have also become normal events, while demand for petroleum-based products from fertilizers to plastics continues to grow. But rather than work to reduce reliance on fossil fuels, Republicans in the US Congress and many business leaders have resisted every effort to find solutions. Instead, through tax breaks to oil companies, and a policy characterized since 2008 by such individuals as the Alaskan populist Sarah Palin as "Drill, baby, drill," the United States pursues oil precisely at great risk to the safety and health of all Americans—and precisely in fragile places with great environmental damage.[23]

Shortsighted fiscal, foreign, and domestic policies create the perfect storm for such disasters as the Deepwater Horizon oil rig in the Gulf of Mexico. More will occur given the aggressive sale of leases to oil companies; three-quarters of the 3,500 offshore production facilities in the Gulf are located off the Louisiana coast. Deepwater Horizon was drilling in water one mile deep and tapping oil another two-and-a-half miles below the seabed. A series of companies operated the well, provided the concrete cap on the ocean floor, and handled logistics and safety. Working with inadequate concern for safety, BP (formerly called "British Petroleum," but the name upset consumers) pushed cost and safety shortcuts, and inevitably a massive accident resulted—inevitably, since more than 12,000 oil accidents had been documented in the Gulf in the previous five years.[24] Eleven workers were killed in the explosion, 4.9 million barrels of oil leaked, tons of the material washed up on Louisiana and Mississippi shorelines causing extensive long-term damage to flora, fauna, and livelihoods. Hurricanes, violent storms, and massive waves threaten the Gulf coast even without oil wells anchored into the ocean floor.

Nuclear Hubris

Perhaps the most egregious of these hubristic decisions concerns the construction of nuclear power stations near cities. Utilities do this for two major reasons: they believe that reactors operate safely and reliably which, with a few major exceptions and dozens of minor ones, they generally do; and they keep transmission costs low. The problem is that if there were an accident that required emergency civilian evacuation, it would be impossible to do so. For example, Soviet leaders hesitated thirty-six hours to order evacuation from the Chernobyl disaster, only to a small extent because of confusing and incomplete information from the accident area. Even with three days' lead time for Katrina, emergency evacuation was confused, confusing, and dangerously delayed at the cost of many lives. If nuclear power stations are built, they should be far from populous cities, in seismically stable areas, and only after issues of nuclear waste and spent fuel storage are solved.

The Fukushima nuclear accident rivals Chernobyl for its radiological impact and the stumbling way in which local and national officials came to grips with the problem. The reactor should never have been built where it was. If not on a fault itself, it was built on the coast of an island prone to a probable tsunami. It would have been nearly impossible to build a seawall to prevent the inundation. Emergency equipment failed. It took engineers weeks to bring some semblance of stability to the reactors—and it took far too long to order evacuation and create an exclusion zone.[25] The Japanese utility TEPCO (Tokyo Electric Power Company), totally unprepared for this dreadful accident, dispatched fire trucks and helicopters to pour tons of water onto the exploded nuclear reactors. This water picked up radioactive isotopes and eventually flowed into the sea; more than 400 tons of contaminated water continued to seep into the ocean late in 2014, affecting marine life and endangering the people who live nearby. Government officials hope that a 1.4 kilometer long wall will prevent some of the outflow.[26] In September 2014, workers sued TEPCO for safer work conditions as they continued to struggle with dangerous levels of radioactivity. The effort to deal with safety, especially that of children, is "fumbling": a slow process that leaves parents unconvinced their children's health is being properly monitored, especially since the government has failed to disclose accurately and regularly the level of radiation exposures. In addition, up to 370,000 children will need to be tested for thyroid problems regularly, and in 2013 the government succeeded in testing only 270,000 children.[27]

The Fukushima nuclear accident reveals another side to disaster history: the central involvement of engineers who really ought to know better. Japan sits on or is near a large number of active faults, and earthquakes occur with greater frequency than in many other countries in the world. A tsunami was inevitable in this area of the Pacific rim. Yet plant design did not involve any kind of mitigation or protection from tsunamis. Six pressurized water reactors and assorted spent fuel pools lost all electricity when the 15-meter tsunami flooded the plant, tore down power lines, and destroyed or flooded emergency reactors and pumps to keep the reactor cores cool. Flooding and earthquake damage prevented or hindered repairs for weeks, and full meltdown occurred in reactors 1, 2, and 3. Fuel rods overheated in pools, workers suffered extreme radiation exposure, and the surrounding area remains to a lesser or greater degree contaminated by unhealthy levels of radiation. Throughout the entire accident mitigation process, the government and utility showed incomplete comprehension of what had happened and how to deal with the situation, revising estimates of the severity of the accident upward with some delay; it was clear to citizens with little nuclear knowledge that Fukushima had entered the lexicon of world disaster history along with Mt. Etna, Bhopal, *Exxon Valdez*, and others.

The handheld video camera and telephone camera enable anyone on the scene of some event to record it. The footage of the tsunami in Japan reveals the full power of the wave as it engulfs everything in its path, ripping apart houses and buildings, carrying the bubbling pieces and shards forward, while automobiles and trucks float away, roll over, and tear into pieces. Everything is swept aside by the cascade—and we cannot see but must remember the 20,000 people whom the tsunami swallowed. Videos reveal the malfeasance of engineers who put the reactor in harm's way. The streets are clear of debris, reconstruction is under way and evacuees are moving out of shelters. But millions of people must readjust to levels of ionizing radiation that were—until March—considered abnormal. Fukushima changed the meaning of "normal," "ordinary," and "natural."

Yet the Japanese are not alone here. Nuclear engineers in other countries predictably responded to the accident with the words "it can't happen here." They greeted Three Mile Island and Chernobyl with the same chorus. Engineers are nothing if not consistent; they have built 432 nuclear power plant units with an installed electric net capacity of about 366 gigawatt (GW) in thirty countries, most of them near big cities, none with the possibility of an orderly emergency evacuation, often in beautiful natural settings to underline their belief that reactors are benign, even when they stand on seismic faults.

In the early 1960s representatives of the California utility PG&E announced plans to construct a nuclear power plant in some pristine area on the California coast, apparently seeing a perfect blending of the machine and the pastoral. Not for a minute did they recognize the cognitive disjunction of the "machine in the garden."[28] They settled on Diablo Canyon with a gorgeous view of the Pacific Ocean. Substantial cost overruns in part the result of debates over the safety of building on an active earthquake fault characterized the construction phase of this reactor—and have characterized the construction of every reactor throughout the world. Utility spokespersons initially estimated costs at $400 million for two units, but by 1976 the bill had risen to $1.2 billion. When unit 1 opened on May 7, 1985, and unit 2 opened on March 18, 1987, the total cost of the plant was $5.52 billion. While selecting the site in the early 1960s, it was not until 1969 that geologists discovered the nearby Hosgri earthquake fault. In an October 1981 article, the *San Jose Mercury* revealed the fault was in the ocean only 4,000 meters from the reactors and that PG&E "knew about the fault for at least a year before telling the public and the Atomic Energy Commission." According to a US Geological Survey report, the station's seismic design could not withstand the maximum potential quake, and this led to retrofitting and upgrading. The Nuclear Regulatory Commission (NRC) licensed the facility after redesign, and the utility has now applied to the NRC to renew operating licenses to 2044 and 2045.[29] The response of public relations representatives at Diablo Canyon to Fukushima was that their reactors were nearly 30 meters above the ocean, with a facility designed to withstand a 7.5 quake in a 6.5 zone, with stored fresh water and diesel generators for emergency operation and cooling. But what if the nondesign 9.0 or 10.0 quake occurs? The 1979 movie *The China Syndrome*, released just two weeks before the accident at Three Mile Island Nuclear Power Station, suggests that profit-seeking utilities may not fully understand the complexity of reactors and what might happen in a meltdown accident at Diablo Canyon or elsewhere.

Soviet nuclear engineers and operators, well-known for their infamous Chernobyl disaster, have a long history of foolhardy and risky station designs, including locating a reactor near an active fault. The Metsamor station in Armenia consists of two VVER-440 reactors, neither of which has primary containment structures. Metsamor lies on some of Earth's most earthquake-prone terrain, and only 30 miles from Armenia's capital, Yerevan. The first unit of the reactor did not involve a seismic resistance system, while the second was designed to withstand an earthquake registering up to 8.0 on the Richter scale. In the aftermath of Chernobyl and the Spitak earthquake both units were closed. But this put Armenia at the energy mercy of its neighbors, and when cut off from natural gas from a pipeline from Turkmenistan during conflict with Azerbaijan, Armenia determined to restart unit 2, a decision supported by the pronuclear International Atomic Energy Agency in 1993. Of course, IAEA nuclear engineers claim it to be "safe." But what of the objections of neighbors Azerbaijan, Georgia, and Turkey to reopening the plant? What of the objections of the EU and the United States, which opposes the reactor restart and denied funds to upgrading it since the reactor has no containment?[30] The EU has declared Metsamor to be the "oldest and least reliable model among 66 nuclear reactors built in Eastern European and former Soviet countries."[31]

This is an interesting case where Russia and Armenia see nucleus to nucleus. Sergei Novikov, a spokesman for Russia's RosAtom, the state corporation interested in seeing its business and reactors spread around the world, argued after Fukushima that Metsamor is perfectly safe. It sits on 120 or 130 hydraulic shock absorbers that ensure the stability of the concrete foundation pad. He said, "When the Spitak earthquake happened, the epicenter had a magnitude of 9, while at the plant site, it was about 7; that is as much as it was in Japan. First the nuclear power plant ceased operations itself, then it started to work mechanically, becoming almost the only source of electricity in devastated Soviet Armenia and then worked for several years."[32] Might we ask Mr. Novikov, what if Metsamor shudders after a 9 on the Richter scale?

I remain convinced that nuclear power has not been regulated as needed, there will be another accident, and we shall have the misfortune to add a fourth name to the now well-known list of Three Mile Island, Chernobyl, and Fukushima.[33] Even at TMI, was the accident always in control? Did officials do what was needed, as needed, to avoid a meltdown and explosion? How close were we to unmitigated disaster and the need to evacuate thousands of people from their homes—and for how long?

We must remember that all nuclear technologies have come from the bomb and they are based on fission processes central to the bomb, that atomic energy is not as cheap as its supporters claim, it never will be, and costs continue to rise into the range of $6–7 billion for one 1,000 megawatt reactor, that radioactive waste continues to accumulate, and that the nations that want to peddle reactors around the world—like Russia, the United States, and France—continue to do so with hubristic confidence that Chernobyl was an error by station operators, not a fault with the technology itself, and will hurry to build where they can.

Even Forest Fires Are Now Human

I looked on in horror in summer 2010 as uncontrolled fires hit central Russia and the government declared a state of emergency in seven provinces. Toxic smoke blew through major cities including Moscow, crops failed, damage topped $15 billion, and who knows how many of the 56,000 excess deaths in the summer heat wave were from smoke and other effects? Were the fires a natural disaster, triggered by an especially hot summer (and global warming)? In fact, a tragedy of human errors combined with Soviet natural transformation projects and post-Soviet greed to trigger the fires. Swamps and bogs surrounding Moscow were drained in the 1960s for agricultural use, afforestation, and to mine peat for power plants. In 2002, a series of hard-to-extinguish peat fires led officials to recognize that peat fields needed to be re-saturated with water to prevent future wildfires. But this never occurred, and many of these dry "bogs" caught fire again in 2010. The result was deep-seated fires that were difficult to extinguish.[34]

Officials claimed the problem was that they could not have anticipated the terrible summer heat wave. In fact, on top of the failure to re-water large regions of peat, as a money-saving measure, in 2007 Putin eliminated the Russian State Forest Fire Service with the assumption that owners or renters of lands would spend the money necessary to prevent forest fires. However, companies that made money on real estate developments had no interest in spending on fire-fighting technologies, only in quick profits. No one—and no entity—was prepared to fight fires.[35] The still-existing Soviet system was based on the long-standing principle that "the earlier a fire is discovered, the smaller the resources needed to put it out." This national monitoring system had thousands of people on the ground, which was extremely effective and cost "tens" or even "hundreds" of times less than the satellite and aerial monitoring the new system required. The new system would have required investment with beneficiaries likely in companies with close ties to the government. Further, this system often fails to detect small fires early on.

The 2007 law was rushed through the Russian parliament (Duma) at the behest of powerful logging interests. This enabled large companies to harvest forests for quick profits, for the law abolished Russia's 70,000 forest guards, who had watched over the trees and called in firefighters, thus making it possible to reclassify forest as lucrative development land. Putin's United Russia party simply sought to enable the state to exploit timber as it had oil and gas—with minimal regulation and with the purpose of short-term profits. Russia's largest timber processing firm, the Ilim Group, was one of the main driving forces behind the new Forest Code, which also benefited the US International Paper Company which owned a 50 percent stake in Ilim. President Dmitrii Medvedev had once headed Ilim's legal department.[36]

Similarly, in the United States (and elsewhere), efforts to manage the forest scientifically and to suppress fires actually increased the scale and intensity of forest fires and wildfires. Whether of human origin (which most fires are, through carelessness, negligence, or criminal behavior [arson]) or from lightning, forest fires destroy vast swaths of land, burn down homes, and kill wildlife and people. Public poster campaigns sought to educate people on these matters.[37] In a word, scientific and economic pressures led to fire suppression, the former pressure based on the belief that humans could regulate natural fires, the latter based on the understandable belief that settlements and towns that increasingly encroach on forests and prairies must be protected. In the United States alone, some 70,000–80,000 wildfires occur annually—with great loss of property and life.[38]

Yet wildfires are a natural occurrence that is part of forest ecology. According to one website, "Many tree species have evolved to take advantage of fire, and periodic burns can contribute to overall forest health. Fires typically move through burning lower branches and clearing dead wood from the forest floor which kick-starts regeneration by providing ideal growing conditions. It also improves floor habitat for many species that prefer relatively open spaces." Suppression permits large amounts of underbrush to accumulate, and when fire does break out, there is much more fuel to feed it. Also, trees that flourish become densely packed, while some such as oak and pine cannot regenerate because they need fire to crack their seeds.[39]

While in the 1930s some forest specialists began to argue for a return to more natural fire regimes, it took massive fires in Yellowstone National Park and great western fires of 2000 to shift perspectives back to seeing fire as natural in fire-dependent ecosystems to reduce the risk of catastrophic uncontrolled wildfire. This required rebuilding of understandings and worldview, and in fact a return to better local understandings of fire that scientists had rejected.[40] The Pacific Northwest has the only rain forest in North America, the Hoh. The Hoh Rain Forest is located in the stretch of the Pacific Northwest rain forest which once spanned the Pacific coast from southeastern Alaska to the central coast of California.[41] Fire has been central to the natural and economic history of the region. According to one source, "Fire was once a natural part of the environment, and Native Americans used it in their quest for survival. But settlers and their descendants regarded fire as the enemy of the forests that generated so many jobs and that symbolized the Evergreen State [Washington]." Again policy makers and foresters adopted fire suppression as policy, a policy that lasted until the 1970s when foresters showed that "suppression had actually been detrimental to forest health and productivity." They had to pursue prescribed burning to eliminate much of the accumulated brush that would eventually feed massive forest fires if not eliminated.[42] This is not the place to discuss the destruction of the Brazilian rain forests—or any other remaining forests. But we see the same mix of political and economic pressure, class difference, arson, slash and burn, and natural and human disasters coming together in landslides, erosion, destruction of ecosystems, alteration of weather patterns.[43]

And we do not learn. We see firefighters attempting to staunch infernos in Southern California that race up canyons with regular frequency. Mudslides follow the infernos some years later. Then the chaparral in the canyons grows back—naturally, becomes a fire hazard—naturally, and wealthy Los Angelenos undertake expensive and extensive efforts to prevent fire and mudslides that are doomed to failure—naturally. They work against a nearly annual natural weather cycle involving the Santa Ana winds.[44]

Anthropogenic Chronic Disasters: Dams

The larger and larger hydroelectric power stations that characterize the second half of the twentieth century require the impoundment of massive quantities of water and the creation of reservoirs of 1,000, 2,000, 3,000 square kilometers of surface area and more. The water inundates fertile farmland and forest and interrupts water regimes, microclimates, and ecosystems; planners frequently undervalue this land when touting the advantages of electricity, flood control, improved transport, and the development of new farming regions through irrigation. They promise to relocate local residents—"oustees"—to new towns with schools, shops, and housing better than before. But this is rarely the case. At the Three Gorges Dam on the Yangtze River in China, 1.5 to 2.0 million peasants have been "removed," and the quality of their lives has noticeably deteriorated. In the Narmada River valley in India, the Sardar Sarovar Dam forced at least 1 million people to move due to the reservoir, canal system, and other allied projects. These mostly poor people will lose their homes, lifestyle, and cultural heritage under water.[45] Jawaharlal Nehru said that hydroelectric and other power stations were the "temples" of modern India. In *The Cost of Living*, Arundhati Roy sees them as weapons of mass of destruction:

> Big dams are to a nation's "development" what nuclear bombs are to its military arsenal. They're both weapons of mass destruction. They're both weapons governments use to control their own people. Both twentieth century emblems that mark a point in time when human intelligence has outstripped its own instinct for survival. They're both malignant indications of civilization turning upon itself.

They represent the severing of the link, not just the link—the *understanding*—between human beings and the planet they live on. They scramble the intelligence that connects eggs to hens, milk to cows, food to forests, water to rivers, air to life and the earth to human existence.[46]

In the case of big hydroelectricity, the anthropogenic costs may never be recouped.

Huge dams require powerful people and organizations to gain the funding and political power to push local people aside in the name of progress. The dams benefit the powerful, the wealthy, and their constituents in urban centers, while enabling the development of resources—lumber, aluminum, iron, fish, and so on—far away. Their supporters tout them for producing copious amounts of electricity, seemingly without generating greenhouse gases (except in the decay of organic material in the reservoir), facilitating flood control, expanding recreational and fishing opportunities, improving transport, and enabling irrigation. But the human and environmental impacts are long-lived and usually irreversible. These include: inundation of agricultural land; preventing fish migration; loss of biodiversity; hydrological changes in water chemistry and temperature; creation of new and poorly understood sediment transport regimes; eutrophication, salinization, siltification; the spreading of waterborne diseases; the loss of cultural and historical heritages; and the social impacts of relocation of "oustees."

Yet the projects move inevitably ahead, supported by such huge quasi-public/private organizations as the Army Corps of Engineers and the Tennessee Valley Authority in the United States; Gidroproekt in Russia; and Electrobras in Brazil. Centrais Elétricas Brasileiras (Eletrobras) sought everywhere and always to expand power generation on the foundation of unquestioned scientific studies. It commands research institutes, utilities, distribution networks, and hydropower stations. Eletrobras has installed capacity of 42,080 MW and 58,361 km of transmission lines.[47] One of its charter projects was the Tucuruí hydroelectric power station, the first major such project in the Amazon rain forest, that dates to the 1970s. It produces 8,370 MW of electricity. The station and dam stretch over 12 km. And its environmental and social costs have been great indeed.[48]

This is especially true of indigenous people who have suffered egregiously at the hands of conquerors and settlers and their microbes, plants, and technologies of control, power, agriculture, and transport.[49] Should we not also consider the destruction of human cultural heritage a disaster? Of the approximately 5 million indigenous people in Brazil, perhaps 200,000 remain. The Xingu people have succumbed to modernization as nature control and hydroelectricity entered their homeland. Elsewhere, as a satellite photo of the Tocantins River and Tucuruí hydroelectric power station reveals in sharp detail, the promise to prevent further encroachment of civilization on the Parakanã Indians is giving way gradually to highways, iron operations, lumbering, and agriculture that inevitably enter the forest along "corridors of modernization."[50] The Tucuruí Hydropower Complex itself flooded 38,700 hectares of the Parakanã Indigenous Reserve. The corridors bring parasites, heavy metals, misguided ecotourism, and so on.[51]

Theoretical Issues: How Has Risk Changed in the Past Century? Do Democracies Handle It Well?

Democracies have made progress in recognizing the risks of life in industrial societies that rely so heavily on large-scale technological systems like highways, tankers, reactors, and the like. Unlike in the former socialist world of the USSR, North Korea, and Eastern Europe—and too often in Russia and China today, where workers toiled without safety clothing, helmets, gloves, steel-toed shoes as they worked with dangerous machinery—a series of safety administrations provide oversight to ensure a high degree of safety for workers and citizens alike. Traffic has been tamed by traffic lights, signs, speed bumps, and crosswalks to protect the pedestrian. Automobiles have been designed with crash integrity, airbags, antilock braking systems (ABS), and other standard equipment. Other public health programs for safer foods have been a mixed success, but exist to one degree or another, much stronger to be sure in the European Union than in the United States.

Yet as Greg Bankoff, Bill Luckin, and others point out, the industrialized nations seem less concerned with the risks that threaten the developing world.[52] They often export their dangerous processes—or hazardous wastes—to those nations, and look the other way as multinational corporations permit their goods to be produced in factories, mills, and buildings without safety regulations, inspections, and other protections, and where workers are under constant threat and pay with permanent injuries or their lives; consider, for example, the collapse of the Savar Building like a house of cards in Bangladesh in April 2013, with the loss of 1,129 lives and more than 2,500 people injured.[53] It was a "Bhopal" in a different form.

Ulrich Beck wrote that Chernobyl was a "shock" because it challenged our understanding of the relationship between technology, society, modernity, and risk. He suggested that industrialized modernity and the wealth it produced was accompanied by the production of risks. These pollution, toxic, radiation, and associated risks were normal occurrences and everyday experiences that officials and specialists had sanitized with Orwellian language. The risks were part of consumption, in food, clothing, home furnishings (e.g., chemicals in "non-staining" fabrics and rugs), and in travel (tens of thousands of deaths annually in the United States, for example), although laws, regulations, and standards attempted to control them. But whereas before people assumed these risks might have a technological solution, Beck argued the risks themselves had become essential to modernity. While societies are more dependent on technology, growing awareness of risk has led other groups to advance rational arguments about the dangers of the modern world in the forms of nongovernmental organizations, lay expertise, and so on. "Science" no longer has a monopoly on risk and its evaluation, no matter how hard technocrats attempt to spin the accidents and dangers modern citizens face. The result is a reflexive critique of risk, and what he calls "reflexive modernization" where multiple sources of knowledge are validated.[54] In addition, as Beck points out, risk no longer respects national borders or class distinctions; Chernobyl's radioactivity spread through the Northern Hemisphere. Still, as the study of environmental racism has shown, risk is not distributed equally in modern society; people of color, workers in such industries as mining, and others bear the burden.[55]

Colonialism, notions of "good governance," and Cold War politics led Greg Bankoff to the conclusion that "there are no such things as 'natural disasters.'" Disasters require human systems in which some people are at greater risk than others because of their place in society. Why then do we use the term "natural disaster" so flippantly, even when large numbers of people lose their lives and so much of their property is destroyed? The problem is that "it suits some people to explain" that natural disasters are nobody's fault. For one thing, they usually occur outside of the developed world. According to the International Federation of Red Cross and Red Crescent Societies (IFRCRCS), "medium and low human development countries (as measured by UNDPs Human Development Index) accounted for 64 percent of the total number of natural disasters, 92 percent of all those reported killed in such events and 97 percent of all those reported affected by them." Bankoff continues that precisely these countries "lack the technological resources to mitigate or even prevent their occurrence." He adds that the notion of natural disasters is in colonial relief and development aid programs and in the help that progressive European nations capable of good governance gave to the less fortunate. But these programs can be ungenerous and have relied on free-market solutions, with the result that the programs served "the 'relievers' rather than the 'relieved.' "[56]

Aid programs, especially during the Cold War, were intended to prevent the aided nation from turning to communism and also to enable penetration of US capital into new markets, with trade relations favorable to the United States. Funding was also often dependent "on the privatization of public services and infrastructure. The resultant privatization programs and sell-off of state assets have commonly taken place [in these nations] in the absence of proper regulatory safeguards, placing many services beyond the reach of the poor, leaving others at the mercy of substantial rises in utility charges, and rendering all more vulnerable to the impact and effect of natural hazards." The poor, for their part, follow through in response to epidemic disease, rapid urbanization, wars, foreign indebtedness, and structural adjustment programs by "living in hazardous locations, inhabiting makeshift dwellings, engaging in dangerous livelihoods or not having sufficient food entitlements." And since "development" translates to large-scale technological systems like mines, dams, and plantations, the result is further dislocation, "ousting" of the poor from their homes and lifestyles into very vulnerable situations. As Bankoff notes, "Those countries experiencing the most rapid rates of growth—much of it based on investment in 'aggressive development' of mega-projects—accounted for 53 percent of all disasters, 64 percent of those killed and 90 percent of those affected between 1999 and 2008."[57]

The Interconnectedness of Natural and Anthropogenic Disasters

I have argued that human settling, engineering hubris, and domestic economic and political desiderata shape "natural disasters." But how does the response to natural disasters differ under "democratic" and "authoritarian" regimes? This topic deserves greater consideration, but it seems that authoritarian regimes respond to disasters more slowly and incompletely. Until Gorbachev, Soviet leaders rarely showed their faces at calamities, and they ordered the media not to provide coverage of natural disasters. They strove to foster the dependency of the citizenry on the government, to instil passivity and complacency, and to prevent the creation of civic culture. They limited the power and resources of local organizations. If the halting and uncertain response of the government to Chernobyl is any indication, then the reputed efficient, highly centralized command and control structure of Soviet power failed miserably. The central authorities were confused by reports from Chernobyl, and they tragically ordered evacuation of the local residents with great delay. Men and women fought heroically to control the accident, but without the proper equipment, let alone with proper medicines. There were no iodine pills to distribute among the population as a hedge against thyroid cancers. Yet Hurricane Katrina and the Fukushima disaster are reminders that most disasters have a central human component independent of polity—by virtue of their proximity, involvement in establishing preconditions, or willful ignorance or malfeasance on the part of political and engineering elites.

Having looked at earthquakes and fires, hurricanes and floods, and having considered the place of human hands and devices in shaping, limiting, or exacerbating these events, there remain two issues to discuss. The first concerns the nature of technological systems embraced in the twentieth century that largely makes negative interaction between nature and society more likely. Highways, levees, hydroelectric power stations, nuclear reactors—all of these things are not only large scale, they also have much greater social and environmental impact when built, and they are more unforgiving should they fail. It is almost a tautology that the negative consequences of a large-scale technological system will almost always be irreversible, and will certainly be recognized only with delay.

From the point of view of disaster history, these large-scale human systems also present threats that are large scale: a reactor meltdown, a ruptured levee or dam, an explosion, collapsing buildings, forest fires, and so on, with or without the assistance of natural forces. Is timely and safe evacuation of large numbers of people even possible in the face of Mother Nature? Unfortunately, often it is not, either because of failures of political systems (late evacuation notices, poor preparedness, underfunded or incompetent emergency personnel, uncaring leaders), or because of the nature of the technology. Engineers have operated with great and understandable hubris in developing technologies to produce electricity, irrigate land and increase agricultural production, house large numbers of people, and protect them in times of crises. But they approach these tasks with self-confidence bordering on arrogance. They believe that accidents with a given technology will not be repeated because they are observant, they learn, and they have engineered failure away. They believed until the 1930s that the levee system would prevent flooding, because during a flood excess water would scour the river bottom and deepen the channel. They have claimed for fifty years that they will produce inherently safe reactors, solve the problem of radioactive waste, and preclude any future catastrophes like those at Three Mile Island, Chernobyl, and Fukushima. They believe there are technological solutions for problems of technological origin. They are certain they will find a fix. And this has created the conditions for a series of natural disasters.

In 2011, terrifying tornadoes tore through Alabama in April with 62 dead and Joplin, Missouri in May (with property damage of $3 billion and 162 dead), while in late August Hurricane Irene soaked the East Coast of the United States. The seemingly growing intensity of these storms and other weather phenomena indicate that, in ways not yet fully clear, human-caused global warming will raise the specter of more and more frequent natural disasters, including more intense disease vectors. Insurance companies have not dismissed the phenomenon of global warming but have hired specialists to study it in an effort to forecast trends and limit losses. If the polar ice caps continue to melt, if holes further open in the ozone layer of the atmosphere, then at the least there will be increased flooding at the coasts including those of islands, and greater numbers of extinctions and cancers. We seem fated to ignore the interconnectedness of politics, science, and nature as the cause of natural disasters.

Vladimir Putin at a wreath-laying ceremony for the *Prince Vladimir* nuclear-powered submarine, Sevmash Shipyard, Severodvinsk, July 30, 2012. For centuries technology has been used to solve practical problems as well as symbolize power, wealth, privilege, and cultural superiority, as this photograph of the proud Putin indicates. Photo by Vladimir Larionov, Severodvinsk, Russia.

6

Big Artifacts

Technological Symbolism and State Power

Technologies of state power obviously include offensive and defensive weapons. Almost everyone is familiar with fighter jets and bombers; destroyers and aircraft carriers; and rockets. Yet even before the twentieth century, big technology served as a symbol of state power, wealth, political authority, or a combination of the three. In medieval towns, cathedrals, then clocks, were signs of prosperity that were visible from the surrounding countryside. The cathedral, a magnificent achievement of geometry, strength of materials, and construction know-how, was the house of God and confirmation of the infinite goodness of the church, a center of civic pride, a destination for pilgrimage, and a place to display artifacts. Gothic architecture represented the authority of the medieval ruling elite, their power, wealth, and proximity to God, and was intended to suggest the awe and admiration of citizens.[1] The clock tower indicated prosperity, perhaps yet to come. In the nineteenth century technological expositions celebrated the joining of state and economic power. The "Great Exhibition of the Works of Industry of all Nations," often referred to as the Crystal Palace Exhibition, held in Hyde Park, London, in 1851, catalogued the achievements of mercantilistic European powers, while the 1876 Centennial Exposition in Philadelphia celebrated not only the Declaration of Independence but "Arts, Manufactures and Products of the Soil and Mine."[2] These expositions demonstrated, in the minds of their promoters and those of the throngs of curious crowds, the epitome of advanced civilization, the power of its industry, and the legitimacy of the political regimes that created such things.

Large-scale technological systems became paradigmatic in the twentieth century as symbols of state power. The monumentalism of the National Socialist Third Reich was intended to demonstrate the racial superiority of the nation and its unassailable power in physical structures that would last a thousand years. Adolf Hitler's architect, Albert Speer, designed parade grounds for Nuremburg spread over 16 square kilometers that, although never built, included a stadium for 400,000 people. Joseph Stalin ordered seven major skyscrapers built around Moscow in the late 1940s to confirm the glory of his rule while people still lived in rubble left from the World War. In 1948 he approved the Stalinist Plan for the Transformation of Nature to subjugate nature itself to Stalinist grandeur through canals, hydroelectric power stations, multimillion-hectare irrigation systems, and forest defense belts. Brasilia in the center of Amazonia, built under President Juscelino Kubitschek, a man of planning and development in the late 1950s, served as a symbol of technocratic rule and was intended to indicate the power of the state to open the nation's rich interior to exploitation, modernity, national destiny in the hinterlands, and freedom from colonial past.[3]

Not only authoritarian regimes have recognized the ideological role of large-scale technologies. On May 25, 1961, President John F. Kennedy addressed a joint session of the US Congress calling for a very expensive and risky effort to put a man on the moon. He said, "[The moon's] conquest deserves the best of all mankind, and its opportunity for peaceful cooperation may never come again. But why, some say, the moon? Why choose this as our goal? And they may well ask why climb the highest mountain? Why, 35 years ago, fly the Atlantic? . . . We choose to go to the moon in this decade and do the other things, not because they are easy, but because they are hard, because that goal will serve to organize and measure the best of our energies and skills, because that challenge is one that we are willing to accept, one we are unwilling to postpone, and one which we intend to win, and the others, too."[4]

The genesis of large-scale technological systems in the complex interaction of economic, cultural, and political forces has been studied extensively.[5] Their ideological significance for nation and state also has served as the focus of analysis, for example, of the development of atomic energy in postwar France or the space race in the United States, the USSR, and in Europe.[6] Whether big technology is the most efficient way to accomplish some specific end has provoked debate. Achievements in big technology distract attention and budgets from social and political problems at the same time as they engender national pride.

Yet most leaders, engineers, and citizens unquestioningly embrace big technology for economic, military, and other purposes, and as icons of national achievement—in the form of modern highways in Germany or the United States, hydroelectric power stations in India and Brazil, rockets and nuclear weapons for North Korea, and the industrial transformation in China.[7] This ideological significance might be called the "display value" of technology, that is, its cultural meaning beyond its technical importance.[8]

Not surprisingly, the government of the Russian Federation under President Vladimir Putin determined early in the twenty-first century to allocate extensive resources to large-scale technologies to shore up the nation's image and self-understanding as a superpower following the psychological shock of the breakup of the USSR. In addition to the military and economic benefits of big technology which Putin and his advisers underline, including annexation of Ukrainian territory in 2014 and reestablishment of army bases in extreme northern latitudes, they recognize the display value of these technologies to secure Russia's place among the leading scientific powers of the world and channel the thinking of the citizenry away from concerns about the present and political dissent and toward feelings of love for the motherland. Combined with state-sponsored programs to develop natural and mineral resources (timber, oil, gas, nickel, platinum, copper, and so on), Putin believes that big technologies indicate the success of his rule and provide the justification for tightening political power over any remaining opposition.

What is surprising is the similarity in the rhetoric surrounding Putin's various programs with those of the Stalin era, and even Putin's unabashedly direct reference to Stalinist programs and approaches to justify investment in the Great Northern Sea Route and Arctic conquest; the military-industrial complex, space, and jets; nuclear power; a kind of Kremlin silicon valley; and even skyscrapers and other extravagant displays of state power, many of whose roots date to the Stalin period.[9] What has been the role of big technology under Stalin and Putin?

Harkening to the Past: The Great Northern Sea Route

In the 1930s Joseph Stalin provided extensive financial resources, personnel, and such new technologies as modern icebreakers to underwrite the effort to secure the Soviet Arctic from Murmansk on the Barents Sea near Norway to Vladivostok in the Pacific Ocean. Scientists, engineers, and explorers journeyed northward at great personal risk, but in their widely published memoirs and public appearances noted their belief that Stalin personally was looking out for them.[10] Like cosmonauts and astronauts decades later, heroic pilots flew a series of bold missions—in this case over the North Pole—to demonstrate Soviet prowess.[11] Explorers wintered on the Arctic ice and studied ice regime, ocean currents, and water chemistry. Communist Party officials worked with leading specialists to establish an entire Arctic empire bureaucracy: the Main Administration for the Northern Sea Route responsible for Arctic economic development whose rapid growth accompanied ambitious national industrialization and militarization programs under the first five-year plans (1929–1941).

Powerful new icebreakers were the essential tool. With them, the "industrialization of the north" would follow. On the eve of the revolution, the Russians had some twenty icebreaking and ice-strengthened vessels operating in Arctic waters. Until the 1930s most of the Russian and Soviet icebreakers that came from British or other European shipyards were underpowered, and it was difficult to get parts for them. By the late 1920s the shipbuilding industry had recovered sufficiently to embark on modernization. They built motorboats for rivers, military vessels, lighters, freighters and icebreakers, although many vessels relied on coal power, which presented serious logistical problems.[12] The technological lag on the Northern Sea Route created significant challenges. Because of growing recognition of the limits of the fleet, in the mid-1930s the Soviet government determined to build new, more powerful icebreakers. Of course, the first ship to be launched was the *Joseph Stalin* in 1938, although two years later than planned. The Ordzhonikidze Shipbuilding Factory in Leningrad launched four ships of the "Stalin" class with length, about 107 m; breadth, 23 m; draft, 9.2 m; displacement, 11,200 tons; speed, 15.3 knots. They could navigate through ice almost 1.0 m thick.[13] In the postwar years, the USSR maintained its lead in and expanded on icebreaker technology with larger vessels, and eventually with nuclear-powered ships.

Stretching roughly halfway around the world, the Russian Arctic covers nine time zones from Norway to the Bering Strait. Approximately one-fifth of the Russian landmass is north of the Arctic Circle. Of 14 million square kilometers that comprise the entire Arctic region (along with the landmass of Canada, the Scandinavian countries, and the United States), Russia's share is roughly 3.5 million square kilometers, one-quarter of the total. (Canada has the largest arctic landmass.) Like Lenin and the early Bolsheviks who saw the Arctic in strategic terms, and after Allied intervention in World War I worried about invasion, Vladimir Putin saw these vast Arctic spaces through military and economic lenses. In a manner reminiscent of the Stalinist 1930s in terms of economic importance, hubristic plans, political legitimacy, and even his rhetoric, Putin reenergized Arctic exploration. Welcoming global warming as an opportunity to develop oil, gas, and other resources, his officials and Russian specialists saw the expansion of the Northern Sea Route as a key to the nation's future. Russia's industrial policy of pushing economic growth on the basis of extraction of raw materials was central to Putin's worldview. He has long believed the development of natural resources was crucial to economic growth and rebuilding Russia's status as a superpower.[14]

Symbolism and rubles have combined to secure the Russian Arctic. Reminiscent of the race between the Soviet Union and the United States to put a man— and flag—on the moon, in August 2007, Russian parliamentarian and explorer Artur Chilingarov engaged in what some observers called a publicity stunt by planting a Russian flag on the bottom of the Arctic Ocean at the North Pole. The government supported the expensive expedition as part of the Russian contribution to the Third International Polar Year (2007–2008). All of the components of Russia's quest for strategic advantage, economic growth, and superpower symbolism were present. A nuclear-powered ice breaker, *Rossiia*, cleared the way for a research ship, *Akademik Fedorov*, staffed by approximately 130 scientists, to get into position for Chilingarov's descent. President Putin welcomed Chilingarov's flag-planting expedition as confirmation of Russia's claim of the Lomonosov Ridge to extend its exclusive economic zone toward the North Pole and several vast oil and mineral deposits. Putin noted that Russia's distinguished history was closely linked to Arctic exploration. Tying these Russian efforts to the great power status of the USSR, he referred to Soviet efforts to build major facilities and cities in circumpolar regions and to the Northern Sea Route in the 1930s.[15]

Russian lawmakers, following the lead of the administration, passed legislation that emphasized Russian sovereignty, underlined the crucial economic importance of Arctic resources, and celebrated the symbolic significance of the northern sea route for Russia's great power aspirations in the twenty-first century. According to a 2001 bill that established Russian Maritime Policy through 2020, Russia reasonably asserted "sovereign rights in the exclusive economic zone for exploration." The policy referred to "the increasing importance of the Northern Sea Route for sustainable development of the Russian Federation." Maritime policy established such long-term objectives as "research and development of the Arctic to the development of export-oriented economic sectors, priority social problems," and "the creation of ice-class vessels for shipping, specialized vessels for fishing, research and other specialized fleets," all toward the ends of state defense and resource development.[16] The Russian Federation would invest billions of rubles to develop gas, oil, apatite concentrate, and many strategically significant nonferrous and precious metals (nickel, copper, cobalt, among others) through state and state-private ventures, with funding for infrastructure, military bases, and occasionally housing.[17]

As it had for Joseph Stalin, the 5,000-kilometer Northern Sea Route from Murmansk to Vladivostok along the Arctic Circle assumed mythic scale for twenty-first-century Russian leaders. In June 2010, then President Dimitrii Medvedev called for the modernization of both military and civilian shipbuilding to enable Russia to engage in the "recently toughening competition for Arctic resources."[18] On May 12, 2012, Putin issued an executive order about the need to modernize Russia's military-industrial complex. He referred without irony to the Stalinist legacy of building military industry in the 1930s with his instructions for "developing the Navy, first and foremost in the Arctic areas and in Russia's Far East with the aim of protecting the Russian Federation's strategic interests."[19] Stalin had developed the Arctic and Soviet industry far and wide. But at what costs? Were the dubious achievements of Stalin in building a military power in the 1930s worth the murder of half of the Red Army officer corps—50,000 men—arrested and executed at Stalin's orders? The creation of the Gulag labor camps and the millions of innocent citizens who toiled—and perished—in them? The 3 million Ukrainian peasants who starved during the collectivization campaign? The poorly functioning economy, especially its poor innovative capacities?

For Putin, however, the big science and technology of airplanes, satellites, drifting ice research stations, and, crucially, icebreakers, including a third generation of nuclear icebreakers, were key to controlling the Arctic. Russian shipbuilders, administrators, and officials evinced great nostalgia for the Soviet Union which created the world's greatest icebreaker fleet. Russia remains the only country to operate civilian nuclear-powered icebreakers, although the icebreaker fleet has aged considerably, and a number of vessels have reached the end of their service lives. Hence Russians of the Soviet generation reminisce about the *Lenin* icebreaker that was launched in December 1957 and sailed on its first mission in September 1959. In the celebratory exposés of the glorious Soviet heritage, contemporary journalists never mention the dangers involved in the rapid, and perhaps premature, embrace of nuclear icebreakers, but instead emphasize that Russians are a full quarter century ahead of the other nations. At the end of the 1950s "we left the Americans behind and first built a nuclear icebreaker," the chief engineer of the *Lenin* atomic icebreaker recently recalled,[20] ignoring the fact that the *Lenin* had two serious accidents in 1965 and 1967, both of which released significant amounts of radioactivity and led to illegal dumping of wastes and reactors at sea.[21]

On August 17, 2012, the nation observed the twenty-fifth anniversary of the sailing of the icebreaker *Arktika* to the North Pole, the world's first surface vessel to do this. The feat celebrated the Soviet subjugation of the Arctic.[22] *Arktika*, a second-generation nuclear icebreaker, was nostalgically retired on October 3, 2008, after thirty-six years of service. It was the fifth of five nuclear icebreakers built at the Baltic Shipbuilding Yards.[23] The others have reached the end of their service, while construction on the most recent addition to the fleet, the *Fiftieth Anniversary of Victory*, a commemoration of Soviet victory in World War II, commenced in the Soviet era, but the ship was not put to sea until 2007 after twenty years of construction owing to extensive construction problems, including a serious fire.[24]

Russian leaders showed only determination to recapture the ideological glories of the Soviet icebreaker. Russia will spend 37 billion rubles (roughly $1 billion) on its next atomic icebreaker according to a contract signed between the Baltic Shipbuilding Factory and Rosatomflot, a subdivision of the Russian nuclear ministry, Rosatom. The new icebreaker has the name *Arktika*, which determines its class (size) and historical tie to the past.[25] Andrei Smirnov, the deputy director of Rosatomflot, Russia's civilian nuclear fleet company, argued that icebreakers will give impetus to exploitation of difficult-to-extract fossil fuels, will enable a five- or sixfold increase in shipping along the Northern Sea Route, and called for an entirely new icebreaker fleet. Icebreakers make not only economic sense: Smirnov pointed out that traveling from Kamchatka to Murmansk takes but 7 days, whereas through the Suez Canal it would take 20 or 25 days, and while the northern latitudes had ice, the southern had something more dangerous—pirates—and pirates "cannot exist in the Arctic in principle: they will freeze."[26]

Technological Utopianism: The Rosatom Nuclear Renaissance

Nuclear power, both peaceful and military, is a more modern technology than the icebreaker with the full essence of superpower status. From the 1940s it served as the engine of the Cold War as the United States and the Soviet Union raced to build tens of thousands of the immoral weapons of mass destruction, joined by England, France, China, and later other nations. They also built nuclear-powered submarines and aircraft carriers, and experimented with nuclear airplanes, rockets, and even locomotives. Since the early 1950s, and President Dwight D. Eisenhower's "Atoms for Peace" speech at the United Nations (1953), it was also a source of propaganda competition as the Soviets and Americans, and later other nations, sought to apply the energy of the atom to industrial, agricultural, medical, and especially energy production purposes in massive nuclear reactors. In the embrace of the peaceful atom, nations of the world touted nearly unlimited energy, in fact "electricity too cheap to meter." The attitudes about cheap energy, declining capital costs for construction, and inherent safety of reactors permeated the thoughts of nuclear engineers throughout the world, including in Soviet Russia, where utopian beliefs about the present and future of nuclear power persist, although the record of peaceful—and military—programs is a frightening reminder of the dangers of elevating symbolism above reality.[27]

At the June 2012 AtomExpo exposition in Moscow, representatives of more than two dozen different nuclear companies, all of which grew out of the Soviet Ministry of Middle Machine Building (Minsredmash), met with potential customers to pursue expansion of nuclear sales. While a number of the companies are connected with operations in Kazakhstan and other former Soviet republics, the vast majority were located in Russia, a fact that reflected the resurgence of nuclear power in the twenty-first century under the leadership of Rosatom (the powerful Russian nuclear ministry). According to Rosatom, the industry is gearing up to bring the peaceful atom to overseas markets.[28]

Ten years ago, as part of state-building of the first Putin presidency, the federal government embraced a crash construction program for nuclear power stations as a symbol of Russia's status as a scientific superpower. At that time this was a dream since the industry was in decay. In the fifteen years since Chernobyl, Russia's nuclear establishment had fallen on hard times and reactor construction lagged. The public remained skeptical of nuclear power; a series of exposés filled the newspapers about past accidents and close calls, not to mention the grotesqueries of haphazard waste disposal that spoiled hundreds of square kilometers of land, especially in the Urals region with great and continuing public health costs.[29]

But Putin and Rosatom officials were determined to pursue a self-proclaimed nuclear "renaissance." Rosatom, a quasi-state corporation, operates thirty-two nuclear power reactors (versus 58 in France and 103 in the United States) with the overall installed capacity of 24.2 GW (gigawatts) at ten power stations. In 2014 they accounted for roughly 16 percent of domestic electricity generation, but were concentrated in western Russia. The share of nuclear generation in the European part of the country reached 30 percent, and in the northwest part of the country it reached 37 percent. A subsidiary of Rosatom, Rosenergoatom, is Europe's second largest utility after the French EDF. Nuclear specialists at the Kurchatov Institute for Atomic Energy, where pressurized water reactors (PWRs, in Russian parlance "VVER") and channel-graphite reactors (the Chernobyl-type RBMK) were designed decades ago, forecast 50–60 GW of installed capacity by 2030, that is, the construction and operation of at least 25 new 1,000-MW new reactors over the next eighteen years, a pace of construction, testing, licensing, and power generation never before accomplished in the world. Anything is possible given that the contemporary Russian nuclear industry constitutes a powerful complex of over 250 enterprises and organizations employing more than 250,000 people—three times the number in France. Nuclear power engineering thus resumed its role as a crucial engine of the Russian economy, a sign of energy independence and sales abroad, and of geopolitical virility.[30]

A great deal of continuity exists between Soviet and Russian nuclear programs, even twenty years after the collapse of the USSR. First, while acknowledging at times the high capital costs per kilowatt hour installed capacity, Soviet and Russian engineers claimed that nuclear power was the only alternative to fossil fuels. If not "too cheap to meter," as the world physics community claimed in the 1950s, then nuclear power, which avoids greenhouse gases, will serve modern industrial society into the twenty-second century when peak oil and gas are part of the past.

Second, engineers have long had dreams of adopting standardized designs for reactors as a way to keep costs down, relying ultimately on "serial production" of reactor components, vessels, and plant facilities. According to their thinking, this would also help quality control in the field and lessen the chance of worker error. The French example provided hope; France produces nearly 80 percent of its electricity from standardized PWRs. Yet at each stage, French specialists have had to introduce safety modifications. The price has risen steadily to $6 billion per nuclear power station, and time horizons for construction have hardly diminished from ten years per reactor. Yet Rosatom specialists, like Minsredmash engineers before them, remain convinced that serial reactor production will succeed and costs will drop significantly, with Russia soon building two, three, even four reactors annually, and at costs significantly lower than those in France.

Builders in the Soviet Union never lacked enthusiasm for large projects, nor was it challenging to gain support for them. From Stalin's canals, metallurgical combines, and entire mining cities, built in part with gulag slave labor, and his grandiose 1948 Plan for the Transformation of Nature, also undertaken with slave labor, to Khrushchev's hydroelectric power stations in Siberia and agricultural programs, and to Brezhnev's new trans-Siberian railroad (known by its acronym "BAM"—the Baikal-Amur Magistral), planners, party officials, and managers found it easier to gather workers at huge construction sites, and everyone believed in economies of scale.[31] No surprise, then, that the nuclear industry built a factory, Atommash ("Atomic Machinery") in Volgodonsk on the Volga River in the 1970s to produce annually up to eight pressure vessels and associated equipment serially in a huge foundry a la Henry Ford. Atommash would ship the 1,000 megawatt electric PWRs by barge and railroad to reactor "parks" of up to ten reactors. But rather than maximizing on serial production and low costs, Atommash produced only three reactor vessels in all before a wall of the main foundry building collapsed in the muck. Apparently engineers failed to take into consideration the changed hydrology of soils on the building site brought about by the proximity to the Tsimlianskoe Reservoir.[32] Hubristic engineers have hardly changed since Soviet times; mass production remains the goal in Russia today. Will the engineers remember to carry out accurate surveys and site selection?

In spite of as yet unsolved waste disposal problems connected with the Soviet military and civilian nuclear legacies, Russia continues to embrace a utopian view of nuclear power as a panacea for energy and for geographic and geopolitical concerns. Sergei Kirienko, the head of Rosatom, and other spokesmen have repeatedly referred to a "renaissance" in the industry. According to Kirienko the nuclear renaissance had three components: perfection and modification of existing reactors, development of fast reactors, and eventually the construction of fusion reactors. "Today we have entered the period of large-scale construction of new stations. The problem is to build the entire system in order to put out one new block in one year, and then two [in one year]."[33] Kirienko asserted, not without foundation, that atomic energy had become safer over the previous twenty years as Russia has embraced International Atomic Energy Agency (IAEA) and International Nuclear Event Scale (INES) standards for operation of reactors.[34] Rosatom's public relations operations have become adept at handling public concerns. The annual "Miss Atom" contest seeks to demonstrate a more feminine side of nuclear power; the 2011 winner, Marina Kiriy, was a mother from the Bilibino Station in Chukotka. The station, the northernmost nuclear power plant in the world, well above the Arctic Circle and part of the country's effort to create a nuclear-powered Arctic and to bring a special kind of "fire" to the indigenous Chukchi reindeer and whaling people, consists of four graphite-moderated EPG-6 reactors, related to the RMBK design, each producing 12 MW electric and 62 MW thermal power (heat) that provides 80 percent of the region's electricity.[35]

Igor Kurchatov, father of the Soviet atomic bomb, pursued peaceful programs with vigorous visions of nuclear-powered utopias until his early death in 1960. He and other Soviet delegates astounded the attendees of the first International Conference on the Peaceful Uses of Atomic Energy in Geneva, Switzerland, in 1955, with the presentations on the Soviet peaceful atom, especially the work of Igor Tamm and Andrei Sakharov on controlled thermonuclear synthesis (fusion).[36] At the twentieth Communist Party Congress in Moscow in February 1956, best known for Nikita Khrushchev's speech condemning the murderous excesses of Stalinism, Kurchatov proposed an aggressive program for the commercialization of atomic energy and applications in industry, agriculture, and medicine that have been expanded and adapted to the twenty-first century.[37]

In patterns now being repeated by Rosatom, Soviet engineers gained government support to accelerate the construction of new nuclear power stations in "parks" of PWRs, RBMKs, and breeder reactors. Embracing technological enthusiasm without sufficient consideration of the risks, Soviet nuclear engineers sought to commercialize breeder reactors on the basis of 1,600 MWe units (the BN-1600).[38] Russia remains firmly committed to breeder reactors that have been costly to operate and waylaid by accidents and fires. The Rosatom goal is to bring the BN-800 on line to demonstrate a closed fuel cycle, and to commence serial construction of breeders in the period 2025–2030. The standard reactors will most likely be 800 MW units and located near the Maiak chemical nuclear fuel facility at Cheliabinsk to take advantage of huge plutonium stockpiles.[39] Russian breeders have the endorsement of President Vladimir Putin and the International Atomic Energy Agency because of the closed fuel cycle. On what foundation is unclear, but Putin claimed that fast reactors are "technically quite feasible." Iurii Kazanskii, a physicist involved in the startup of the BN-600 in 1980, called the completion of the BN-800 "a question of the leadership of Russia."[40]

The Northern Sea Route, Arctic gas deposits, and nuclear power have been confidently linked to one of the most troubling technologies to drift from Rosatom Arctic tides—floating nuclear power stations. Based on submarine reactors, and developed in Severodvinsk and other military R & D facilities to maintain high levels of employment in the nuclear shipbuilding industry, float-ing reactors (and floating nuclear-powered oil platforms still in the design stage) will provide electric energy and heat above the Arctic Circle. To achieve this "maritime" Bilibino, Russia has announced plans to build twelve floating nuclear reactors. The reactors will produce 90 MWe, but could also be designed for desalinization and industrial heat production. Rosatom plans to sell them for $335 million each; China, Algeria, Indonesia, Brazil, and other countries have indicated an interest in the plants. The first floating reactor was planned to be operational by the end of 2013, but apparently will not be operational until late in 2015, and will be moored near Petropavlovsk-Kamchatsky. The region is seismically active, but Kirienko dismissed Fukushima as irrelevant to the Russian experience concerns. He said, "I know Fukushima has sparked many inflammatory rumors and gossip, including on the floating nuclear plant. Some people say that if a ground plant could not withstand a tsunami, what would then happen with a waterborne nuclear plant. But nothing will happen. Everything will be just fine."[41] Everything will be fine, according to the inhab-itants of Pevek, in the Arctic Circle, who have apparently welcomed the next floating reactor, whenever it arrives, for its promised heat and electricity.[42] Safety, proliferation, terrorism, tsunamis—all of these things have little place in Rosatom's world when heat and light will be the result.[43]

On March 18, 2010, on a visit to the Volgodonsk nuclear power station to promote the future of the "Russian peaceful atom" and reveal his modern leadership, Putin declared that Russia was prepared to claim one-fourth of the world's reactor market, and not just cap Soviet achievements in the domestic arena. In a distinctly Soviet ceremony, Putin himself pushed the "power" button in the control room and then gave state prizes to "outstanding" atomic workers, perhaps even more outstanding than the "heroes of socialist labor" Commu-nist Party officials celebrated but two decades earlier. He was confident that within a short time frame Rosatom would bring online nearly as many reac-tors as were built during the entire Soviet period. Putin wistfully referred to the unlimited potential of Russian programs: "We are fully capable of taking no less than 25 percent of the world market in construction and operation of AES."[44]

In all of these ways, pride and technological momentum remain central aspects of the Russian nuclear industry since its founding in the 1950s. Nuclear power serves as a panacea for power production, transportation, and uneven distribution of resources and population. It warms and illuminates the masses. Engineers have no doubts they can design safe reactors with a variety of different applications, and use their original purposes interchangeably. They have the support of the president, the state, and a large branch of the economy. Leaders and ministers see nuclear power as a symbol of great power status and support billion-dollar expenditures accordingly.

Missile Envy: The Stalinist Military-Industrial Complex

It is not surprising that nuclear nostalgia triggered military envy. Russian leaders have been increasingly vocal in celebrating Soviet achievements—Stalin's joyous establishment of the nation as a military power in the 1930s, the first satellite, Sputnik, and man, Yuri Gagarin, into space, in 1957 and 1961 respectively, and, of course, various nuclear achievements—with the exception of Chernobyl. Space has been centrally important to Russian self-image and imagination.[45] Gagarin, for example, was a new kind of hero—a hero of the potentialities of Soviet society under Khrushchev, and of reborn faith in the communist future. Under President Putin, Gagarin's heroism has been reborn to serve the state.[46] On the fifty-second anniversary of Yuri Gagarin's flight into space, Putin unveiled a $50 billion drive for Russia to preserve its status in space, including the construction of a new cosmodrome at Vostochny in the Amur region of the Far East.[47] As Stalin forced the nation to "reach and surpass" the West, so Putin announced that Russia will send manned flights from its own soil in 2018 from Vostochny to deep space as well as moon missions as part of the effort to catch up and overcome the gap in "so-called deep space exploration" and for Russia to "preserve its status as a leading space power." Then, following Brezhnev's lead, Putin congratulated cosmonauts on Russia's Space Exploration Day: "These are not just any greetings, these are greetings from the construction site of our future."[48] On April 12, 2014, Putin appeared at the Space Museum in Moscow to celebrate Gagarin Day and announced a plan to colonize and mine the moon.[49]

Russian leaders may well embark on this new space effort through the new "Angara" low-earth orbit rocket. The Angara project dates to the 1990s and represents an ongoing effort to be free from Soviet dependencies on Ukrainian missile construction facilities and Kazakh-based launches from the Baikonur cosmodrome that supported Gagarin and others. Unfortunately, to date, the Angara has yet to lift off successfully from Plesetsk in Arkhangelsk province.[50] Other rocket disasters belabor the industry.

Putin has also pushed Russia to recover the Soviet heritage of military and civilian jets. While the Russian government puts money into high tech, its low-tech infrastructure of mines, roads, railroads, ships, and Soviet-era passenger jets will have a hard time competing abroad. Much of the industrial base dates to the 1960s or earlier.[51] Nonetheless, the authorities have determined to pursue a passenger superjet in the tradition of the Tupolev, Iliushin, and other airplanes whose sardine-like cabins provided legendary discomfort for the passenger. In April 2011 President Medvedev called for Russia to "upgrade the civil aviation fleet. Passenger planes flying the main routes have an average age of 17 years, and regional planes are even older, up to 30 years. These are very old aircraft."[52] Russia can hardly make spare parts for its aging fleet, let alone compete with the European Airbus, the American Boeing, the Canadian Bombardier, and the Brazilian Embraer passenger jets. The Superjet 100, the only commercial airliner designed and built by Russia since the fall of the Soviet Union, was meant to generate sales and "to restore at least some of the prestige that Russian engineering had lost after the Soviet collapse." President Putin supported the Superjet "as a point of national pride."

When big technology fails, it fails in terribly costly and public ways—as the *Challenger*, *Exxon Valdez*, and Bhopal, India, disasters indicate.[53] Since it can no longer control access to news about technological failure as during the Soviet period, the government and its state-controlled high-tech agencies run the risk of public examination of any disaster, whether the sinking of the decrepit *Bulgaria* passenger ferry in the Volga near Kazan in July 2011 with the loss of more than 110 passengers,[54] forest fires in Moscow in 2010, floods in Siberia and the Far East in 2011 and 2014, or a modern Superjet. The crash of a Superjet in Indonesia in May 2012 indicated the problems and pitfalls of seeking state power through high technology. No sooner had the jet crashed than the Russian press published articles in which some unnamed xenophobic individuals suggested that the United States must have contributed to the accident by jamming the jet's communication system.[55] How else could this state program fail without some dastardly outside, ill-intended party acting, they posited, rather than allow for the fact that accidents unfortunately occur and that sometimes Russian engineers are responsible for them—as dozens of space failures, Chernobyl, and other airline accidents indicate.

Problems with the new Sukhoi indicate that the confidence of the Putin administration may be misplaced. In 2013, a Sukhoi Superjet aborted a takeoff from Sheremetevo Airport in Moscow after an engine failure. Aeroflot grounded four of its Superjets in February due to "technical problems," while a flight from Moscow to Kharkiv, Ukraine, was also aborted during takeoff because of engine failure. Sukhoi blamed the problem on Aeroflot maintenance, while Aeroflot refused to comment. But the president insists Russia will compete with Brazil's Embraer and Canada's Bombardier, and claims that Russian industry will sell $250 billion worth of aircraft by 2025 and then compete with US and European giants.[56]

The fight to resurrect the jet as a symbol of post-Soviet verve commenced under Putin with the merging of the Iliushin, Mikoian, Irkut, Tupolev, Iakavlev, and Sukhoi companies in the state-controlled United Aircraft Corporation.[57] In recent years the government has supported this corporation to modernize factories and lower the cost of serial production of airplanes. Russia's Tu-134, Tu-154, Il-62, and even the Il-18 were outdated, expensive to operate, used vast amounts of fuel, and needed repairs that required impossible-to-get parts.[58]

These design bureaus and their factories had a glorious past dating to the 1930s; Stalin's compatriots carried out a campaign to glorify pilots and planes and pursued technological hero worship. But the aerospace industry under Putin faced an uncertain future because of cost overruns, concerns over safety, lags in manufacturing capability, the need to rebuild capacity from design to manufacture, and lack of international interest. The company Antonov tried to introduce the An-148 to compete with the Boeing 737, but it was too expensive to operate. A cargo version, the An-178, had no market. Without a captive market and without the ability to subsidize aircraft as the USSR did during the Cold War, the Russian civil aviation sector may not survive competition with better-made and more modern Airbuses and Boeings—and their worldwide product support.[59] Fully 50 percent of the Russian industry is state-owned, versus 30 percent when Putin assumed power. Like Peter the Great and Joseph Stalin before him, Putin intends the Russian state to lead the way in modern science and technology in partnership with resource development programs directed by close oligarchic advisers. But the barriers to innovation and progress—including political control of capital, a closed political system, and the lack of a culture of innovation—are legion.[60]

Oligarchic Exploitation: The State of Oil and Gazprom

Another facet of technological display in modern societies comes from the combination of corporate strength, state authority, and vast quantities of capital. Many people identify "America" with such corporations as McDonald's, Coca-Cola, and Exxon. Such massive organizations actually date to the beginning of the 1600s and the British East Indies Company that came to control India through its private armies and the Dutch East Indies Company which gained the wealth and power to wage wars, establish colonies, and even imprison and execute convicts. In the nineteenth century they accumulated enough power and wealth to create monopolies, engage in price-fixing, and lead to the call for antitrust legislation. They remain massive and powerful into the twenty-first century; the US Supreme Court even gave them the status of "people" and rights of free speech.[61]

Similarly, there is no shortage of oligarchic state companies in twenty-first-century Russia, whose logos have become recognized throughout the world and whose wealth has few rivals. The most widely recognizable, Gazprom, a Soviet-Russian hybrid, with nearly 400,000 employees, self-consciously em-braced display value.[62] Its logo fills posters throughout the nation, supplanting the once ubiquitous Soviet political posters. Two major artifacts of natural gas reveal both the promise and the challenges of playing up the symbolism of big science and technology in contemporary Russia: the Shtokman fields in the Barents Sea and the Gazprom Skyscraper in St. Petersburg. They make quite clear the unlimited power of industries connected with the Russian presidency.

Launched like the *Stalin*, *Lenin*, and *Arktika* icebreakers with anticipation of great economic benefits, but like many Soviet expeditions caught unexpect-edly in the ice, the Shtokman fields have not opened in spite of the hundreds of millions of dollars thrown their way. Initially owned by Gazprom (51%), Total SA (France, 25%), and Statoil ASA (Norway, 24%), the Shtokman Devel-opment AG had a budget that exceeded $800 million for 2008–2009,[63] based on a promise of access to reserves estimated at more than 4 trillion cubic me-ters of gas; an agreement signed in 2008 anticipated production beginning in 2013–14. But just four years later, the project was frozen. According to the Bellona Foundation, "The announcement from the Russian state gas monopoly indi-cates that even the Russian government cannot, for the time being, see its signature gas project yielding a healthy financial return." This was a disappointment for President Putin and his economic strategy, for Russian leaders saw liquefied natural gas for export as key to the country's financial growth. Both Total and Statoil have given up their shares of the field as too environmentally risky and expensive. It does not help that Russian laws make it difficult for Western companies to join any Russian project,[64] and that original estimates of costs have doubled, to $30 billion.[65] And it hurts that, in response to Russia's annexation of Crimea and war in Ukraine, the European Union and United States have embargoed vital oil equipment for future Arctic exploitation, and one-fifth of Russian oil and gas production is at risk.[66]

But Gazprom insists on being known as a success story with its stony blue-flame logo plastered everywhere. It also plans a grotesque skyscraper that is inappropriate by any standard to blight the St. Petersburg skyline. Like the monumentalist Stalinist skyscrapers and apartment buildings reserved for Soviet elites that demonstrated the omniscient glory of state power, so the planned Gazprom tower indicated that Russia's new elite class of resource developers have exclusive access to the Kremlin. The controversy over the extravagant Gazprom skyscraper reveals the determination of the huge conglomerates to paste their advertisements of wealth not only on billboards and soccer jerseys, but above the skyline, while ignoring the will of citizens to preserve history. Like the seven Stalinist "wedding cake" skyscrapers in Moscow that were symbolic of Stalin's glory, and were built when millions of people lived in *kommunalki* (shared apartments, including sharing kitchens and bathrooms with strangers), in the rubble, and in underground *zemlianki*, they indicate power run amok among oligarchs. Like a Soviet-era ministry with no checks on its programs and with legendary access to manpower and resources, Gazprom directors want to build Europe's tallest skyscraper in St. Petersburg, a city comprised mostly of five- and six-story buildings. Designers claim that the tower will "make St. Petersburg a world city" and be true to the city's history, for example, with a base "modeled on the pentagonal shape of the ancient Swedish fortress Nienshant[z] . . . to pay homage to the myth of St. Petersburg as a city built on water (the spiraling glass structure representing water)."[67]

The first skyscrapers bespoke the wealth of corporate owners, but also celebrated national culture, and capitalist ambition, energy, and enthusiasm. They grew from the Woolworth Building (1913), to the Chrysler and Empire State Buildings of the 1930s, the World Trade Center and Sears Tower, to a veritable skyscraper-building orgy in China and the Burj Khalifa in Dubai, the twenty-first century skyscrapers that include luxury apartments for the wealthy to the glorification of state and economic power. Gazprom owners—the Russian state and Putin's oligarchs—intended the same. Gazprom directors were thrilled with the original plan to erect the structure across from the Smolnyi Institute, a women's school from the tsarist era, Vladimir Lenin's first seat of power, and where Sergei Kirov, first secretary of the Leningrad Party organization, was murdered on Stalin's orders in 1934, to create a business district near the downtown. It is not clear what Lenin would have thought of the juxtaposition of capitalist greed with political power. In any event, the tower's monstrous size was a violation of local sensibilities, commonsense aesthetics, and likely of UNESCO "World Heritage" designation for St. Petersburg. After its approval by Russia's State Expert Evaluation Department, in no way a zoning commission, the St. Petersburg Committee for City Planning and Architecture and the City Court approved the crystal tower. The City Council waived height restrictions to permit the project to go forward, but public outcry about the fact that the crystal shaft would destroy the skyline and distort the historical center of the city slowed the project, and President Dimitrii Medvedev, who served on the Gazprom board until he became president in 2008, announced his opposition. Surprisingly, Gazprom had to reconsider.

Not surprisingly, Gazprom returned with an even taller palace. In August 2012, President Putin made a fast-track decision to approve the new design, the so-called Lakhta Center, on the city's outskirts near the Gulf of Finland. If nothing else this decision indicates the power of the Kremlin to intervene arbitrarily. Gazprom took the decision on the new site two weeks before planned public meetings in order to prevent any input from citizens since the first project had been so roundly shunned. The spire, at 1,640 feet (500 meters), would include stores, restaurants and cafes, and is 300 feet taller than an earlier proposed skyscraper.[68] Gazprom officials and their handler, Mr. Putin, a Petersburg native, apparently find no aesthetic dissonance between an ugly crystalline phallus and the historic, human-scale buildings that fill Petersburg.[69]

While moving the tower out of the center to the northwestern edge of the city near the Gulf of Finland, its construction will lead to traffic, pollution, environmental and other problems,[70] and will fail to create a "business center" in Petersburg, which was one of its stated goals, although a "failed business periphery" is a real possibility. Boris Vishnevsky, a journalist and local legislature member of the opposition Yabloko Party, said, "Even nine kilometres [5.6 miles] from the centre, the building will be the most prominent object that an eye can see."[71] Vishnevsky referred to the excrescence as "Gazoskreb" (Gas-scraper). The question is whether various political parties, Russian NGOs, architects, journalists, and UNESCO have any chance to bring down the Gazoskreb before it goes up with the support of the city, Putin, and Gazprom.[72]

The Lakhta Center will be a traffic magnet and environmental disaster. It involves the construction of a new metro station, of course, but also of a new embankment along the Okhta River; the extension of Sverdlovskaia embankment; the construction of a multilayer modern flyover on Krasnogvardeiskaia Square and a tunnel underneath it; and the widening and extension of streets to the Ring Road. Parking garages and parks of 25,000 square meters (6.2 acres) will draw more smoke-belching vehicles to the region.[73] Petersburg's legendary traffic jams will extend to the Gulf of Finland. But the skyscraper will satisfy the interlocking interests of the state and gas industry for a project visible from the bedrooms of every citizen, reminding him or her where power lies.

Technological Display in Full Flower: Snow in the Black Sea

We have seen so far how, in modern technological societies, state power aspirations, industrial strength, corporate logo-ism, and utopian visions have come together in large-scale systems whose cost may far outweigh their benefits, but which have acquired significant value symbolically for the legitimators of those states, especially in aero-astro, construction, nuclear, exploration, and other technologies. Yet in authoritarian political systems perhaps technological enthusiasm goes even further, because there are few brakes on its embrace by the state. The public may have been emasculated by the police, or perhaps a vital civic culture never really formed, as in the case of Russia.

In the Sochi Winter Olympic Games, as in Hitler's 1936 Berlin games, Putin wedded state power, wealth (here petrorubles), and visions of gold medals—the latter achieved especially on the last day of the Olympics with a sweep of the gold, silver, and bronze medals in the men's 50km ski race—with the goal of demonstrating to the Russian citizenry the recovery of the nation from the embarrassments of the 1990s, including the breakup of the Soviet empire, an allegedly farcical leader, Boris Yeltsin, who drunkenly played into the hands of the American capitalists and CIA, the collapse of the economy, and the shock of a demographic crisis that left Russia the only industrial power with a declining population as deaths exceed births.

The Putin administration was determined to use Sochi to deflect public attention from those problems—and from its assault on personal freedoms and its newly passed homophobic laws—in a grandiose celebration of state power. The celebration required Putin's oligarchs to pay for his leadership. Throwing environmental caution to the warmer winds of the Black Sea resort, the government spent at least $50 billion, much of it going to corrupt projects and individuals, for buildings, stadiums, venues, and snow-making machines to build a winter wonderland of benefit to the wealthy in an inappropriate climate, and with significant environmental degradation the result.[74] Described by one observer as "outsized in scale and ambition," the cost exceeded the 2010 games in Vancouver, Canada, tenfold. A company owned by Vladimir Potanin, one of the wealthiest men in the world whose fortune includes Norilsk Nickel, complained that one of his companies had to pay out $530 million in extra work. Other oligarchs complained as well, but Putin determined that costs were no object because he wanted the Sochi games to project an international image. One of his spokesmen said, "All (rises in costs) there are justified. It is not possible to calculate everything in advance. New demands arise, including those from the International Olympic Committee, which require additional costs. There's nothing extraordinary about it."[75] Hidden not far from the Olympic village, the number of Russian people living in poverty continued to grow, human rights violations played out in the name of protection against terrorism, and activists were arrested, jailed, and sentenced to long prison terms.

The opening ceremony of the Sochi Olympic games was a bizarre self-referential celebration of the Russian past and present that glossed over the authoritarianism of the tsars and the murderous policies of Stalin while incorporating symbols of that Russian past: robot bears; a likeness of the "bronze horsemen" statue of Peter the Great; the Soviet hammer and sickle (as historian Matthew Light wondered, how would people react to a swastika being displayed at an Olympics in Germany?); and castles, fortresses, and churches that reminded viewers that the Russian Orthodox Church has reunited with the Kremlin after the Soviet interregnum, the same church that joined Putin's United Russia political party in trying to dominate elections, happily saw the prosecution of the Pussy Riot rock group, and spearheaded the effort to tie conservative homophobic values to state policies. There were a few glitches: the Olympic torch, developed by a Siberian factory that produced ballistic missile parts, which traveled to the cosmos and back on the way from Athens to Sochi, self-extinguished a number of times and had to be re-lit with cigarette lighters; one of the Olympic rings failed to light during the opening ceremony. But four of five is a passing grade.

Of course, other nations have used the Olympic games to do more than showcase the world's greatest athletes. Adolf Hitler in 1933 instituted an "Aryans-only" policy for the Reich's athletic teams, although he convinced other world and Olympic leaders that his policies would have no impact on the games and avoided a boycott. The Nazi government presented a peaceful image of Nazism during the 1936 Berlin games, but used propaganda to draw parallels between the Reich and the Aryan purity of ancient Greece. So, too, on the eve of the Sochi Olympics, the Putin administration introduced laws enabling the prosecution of gays and lesbians, yet assured the international audience that these laws had no relevance for the Olympics. Three years after the close of the Berlin Olympics the Nazis invaded Poland, and two years later, Ukraine. Putin annexed Ukraine's Crimea three days after the conclusion of the Sochi Olympics. But, like Hitler, Putin himself said little at the Olympic games. Nor did the Russian people, who had been cowed into silence. But he smiled smugly, having achieved his goals.

The symbolism of new Russian imperialism figures as well in the rebuilding of St. Petersburg for the approaching 2018 World Cup Soccer championships. In Petersburg, as in Sochi, crumbling Soviet-era infrastructure must be fixed rapidly and new infrastructure must be built, but the government will focus on ring roads and football stadiums to the detriment of all other needed rebuilding and modernization in provincial Russia.[76] Russia is rebuilding and modernizing soccer stadiums, none more expensive than the new "Zenit" football club stadium at a cost of 44 billion rubles ($1.4 billion, which rivals the monstrously expensive Yankee Stadium in Bronx, New York, built with public funds).[77] Located along the waterfront on the way to the planned Gazprom skyscraper, the project has commenced apparently without contributions from Gazprom, Zenit's majority owner, which means the Kremlin owns the stadium. Yet the public has received the bill, and the price tag has increased sixfold since the cost was first announced, although the size of the stadium increased only 10 percent. Most shocking is the fact that Governor Georgii Poltavchenko of St. Petersburg asked the residents of Petersburg, in particular the fans of Zenit, to consider voluntarily working on the stadium to keep costs down in the good Soviet fashion of getting workers to give up weekend rest days to build socialism by cleaning, pruning, painting, and so on (*subbotniki*, days of "voluntary labor").[78]

Political Power and Technological Conservatism

Beyond ideological functions, large-scale technologies have many useful purposes: flood control, power production, increases in agricultural bounty, job creation. During the New Deal, US president Franklin Delano Roosevelt spoke at several occasions on the importance of modern large-scale, government-supported projects for electrification, expansion of industry, and improving the quality of life of the people of America. Announcing the New Deal in July 1932, Roosevelt called for the "improvement of a vast area of the Tennessee Valley . . . to the comfort and happiness of hundreds of thousands of people and the incident benefits will reach the entire nation."[79] At the dedication of the Grand Coulee Dam on the Columbia River in Washington State in August 1934, Roosevelt spoke about the huge increment to power production for agriculture and industry and to the family home. The dams would enable the "men and women and children" to make "an honest livelihood and [do] their best successfully to live up to the American standard of living and the American standard of citizenship."[80] And at the dedication of the Bonneville Dam in September 1937, Roosevelt identified the dam with "the future of the Nation. Its cost will be returned to the people of the United States many times over in the improvement of navigation and transportation, the cheapening of electric power, and the distribution of this power to hundreds of small communities within a great radius. . . . As I look upon Bonneville Dam today, I cannot help the thought that instead of spending, as some nations do, half their national income in piling up armaments and more armaments for purposes of war, we in America are wiser in using our wealth on projects like this which will give us more wealth, better living and greater happiness for our children."[81]

In rhetoric and reality, scientific and technological prowess remained important as symbols of superpower status, the will of man and woman, the crucial place of indefatigable curiosity, and the inevitability of discovery. On June 26, 2000, US president Bill Clinton celebrated the completion of the mapping of the entire human genome (the Human Genome Initiative, or HGI) in a speech that likened the HGI to conquest of the New World. He said,

> Nearly two centuries ago, in this room, on this floor, Thomas Jefferson and a trusted aide spread out a magnificent map—a map Jefferson had long prayed he would get to see in his lifetime. The aide was Meriwether Lewis and the map was the product of his courageous expedition across the American frontier, all the

way to the Pacific. It was a map that defined the contours and forever expanded the frontiers of our continent and our imagination. Today, the world is joining us here in the East Room to behold a map of even greater significance. We are here to celebrate the completion of the first survey of the entire human genome. Without a doubt, this is the most important, most wondrous map ever produced by humankind.[82]

Clinton celebrated the "combined wisdom of biology, chemistry, physics, engineering, mathematics and computer science" and the force of "more than 1,000 researchers across six nations" in the achievement. He reminded his audience—those present and the American people—that the HGI was "more than just an epic-making triumph of science and reason. After all, when Galileo discovered he could use the tools of mathematics and mechanics to understand the motion of celestial bodies, he felt, in the words of one eminent researcher, 'that he had learned the language in which God created the universe.' "[83]

Never one to mince words, if often to mispronounce them, President George W. Bush also drew on big science and technology to legitimize his presidency. In 2004 he called for NASA "to gain a new foothold on the moon and to prepare for new journeys to the worlds beyond our own." He proposed sending robotic probes to the lunar surface by 2008, a human mission by 2015, "with the goal of living and working there," and then "human missions to Mars and to worlds beyond." He recognized that it would cost billions of dollars, but had no doubts about the worth of the expense.[84]

Under Presidents Vladimir Putin and Medvedev, the Russian government has ventured directly into science and technology policy to preserve Russia's status as a scientific superpower. The government found billions of rubles for continuing space research, resurrecting atomic energy, expanding Arctic programs with a nonpareil nuclear icebreaking fleet, and moving into passenger jet service to complement the sale of military jets. The Putin administration selected these regions of modern technology to demonstrate that the nation had not only significant resource wealth, but also scientific and engineering excellence befitting a superpower. Additionally, Putin has resurrected the symbolic awards and ceremonies of the Soviet era to praise the men whose achievements radiate onto state power, including a reborn "Hero of Labor" prize.[85]

If, under Stalin, the trappings of Soviet gigantomania reflected state power and the omniscient power of the leader, then new Russian technological display grows out of that same state power tied to massive industrial resource combines—Gazprom, Lukoil, Rosatom, and others. The pronouncements of leaders and officials indicate the importance of big technology for contemporary Russia as proof that the nation remains a superpower at the cutting edge of technology, with military might as an added bonus. Space and nuclear power, for all of the risks and costs associated with their pursuit, are at the top of the list of crucial state programs. Perhaps with the blessings of the Orthodox Church Patriarch, the Putin administration can avoid such problems associated with technological symbolism in the Soviet era as technological failure. But the reliance on big technology is problematic in another way: Moscow has become a black hole of power and money, while towns and cities in many regions of provincial Russia have inadequate budgets for repair of infrastructure—roads, bridges, and public transportation. Yet in this environment, patriotic engineers who desire support for grandiose projects have unsurprisingly even rekindled a variant of the on-again, off-again project to "reverse" the flow of Siberian rivers to provide water for industrial and agricultural purposes in Central Asia and to replenish the Aral Sea. The environmentally unsound, grotesquely costly project will cost at least $40 billion and rival the pyramids as symbols of engineering hubris.[86] The question is whether President Putin wants such a canal.

A bicycle park in Amsterdam, The Netherlands. As an alternative to the automobile, the bicycle is a model for using existing technologies for a more reflective and less consumptive way of life. Photo by Vera Kratochvil.

What Have We Learned from This?

Books, Bicycles, and Other Things That Go Bump in the Light

In the nineteenth century, German philosophers debated whether a "thing in itself" (Ding-an-sich) existed, if we could know its essence, and what that essence was. Immanuel Kant suggested that the human mind could never penetrate into the essence of the thing in itself but could perceive that thing—how it appears to the observer. This book took a different approach to the thing in itself. Together we have investigated a series of things, technological artifacts that, at first glance, appeared to be simple things intended to fulfill one purpose, in some cases objects you could hold in your hand, or wear, or eat.

But each of the objects in fact, and not surprisingly, grew out of and reflected a series of cultural, social, political, economic, and environmental relationships. The granules of sugar on your kitchen table grew out of the powerful and immoral institution of slavery and to this day the backbreaking and underpaid labor in the Florida Everglades and elsewhere. They reflected colonial economic relations and later major corporations. Throughout, economic organizations have benefited from direct and indirect subsidies, land grabs, and state-funded reclamation projects to make more land available to the companies. The granules are trade organizations now engaged in legal battle with other trade organizations representing multinational companies that process corn to make an artificial sweetener that is added to many processed foods in America. No matter the links of the foods to rising obesity and increasing frequency of type 2 diabetes in the United States; the needs of schoolchildren for safe food and a balanced diet have been drowned out by food manufacturers who insist that all is well and that the government ought to facilitate lunches based on too much starch and sugar and salt and not enough fresh fruits and vegetables.

Fructose has moved to the European Union where the same battles over health, safety, and schoolchildren may play out. With the assistance of input from corn syrup exporters, the EU published its Regulation 536/2013 that amended a recently published prior regulation and permitted health claims made on fruc-added foods without referring to potential disease risk and children's development and health problems.[1] According to an EU public health NGO, "If the manufacturer replaces 30% or more of the glucose or sucrose in its product with fructose, it can place a health claim on the product stating that it reduces blood-sugar spikes after eating." The result will be "a flood of high-fructose products appearing on the EU market, all proudly proclaiming their health benefits."[2]

Similar individual and bureaucratic, producer and consumer, political and economic concerns come together to make any technological artifact what it is: vested interests, political determinations, and cultural values that shape the speed of its diffusion, organize its productive relations, and determine the environmental costs and other euphemistic "externalities" (usually meaning ignored costs). As historians, students, and curious readers whose lives are affected directly by the objects around us, we can and should ponder the place of any technology. The process may be entertaining, and it may be frightening and upsetting, or it may simply be enlightening. But follow the money to understand an artifact, its genesis and diffusion, its success or failure, the extent to which it becomes a cultural icon like the banana or automobile or soft drink can. Then consider the political and regulatory environment and vested interests in the artifact. And finally, ask what can be done to make technologies less disruptive to our daily lives. I urge thoughtful personal resistance.

Without the luxury of the detailed analysis in earlier chapters, permit me to offer two other examples of objects important in everyday life whose relationships with us reveal truly our relationships with others—with buyers and sellers, policy makers and regulators, workers and managers, engineers and planners, fellow citizens, commuters and road ragers, library directors, and the objects themselves—no matter how distant or innocuous they appear to be. Through historical analysis we may bring them into temporal and almost physical proximity so that we are prepared to see technology for what it is: an inspiring product of human ingenuity and cunning, sometimes also of deceit and naked self-interest, and always of unexpected uses and impacts on our lives.

Books and the Internet

Books and their various homes—the bedside table, the summer hammock, shelves in the study, the backpack, and especially the library—have been the major icon of informed, polemical, and dangerous thoughts for millennia. Libraries housed books that at first were painstakingly handwritten, next produced with moveable print, then by presses, and ultimately in electronic form. Monks not only gave us brandy and bitters, but also modern libraries in the early Middle Ages. Universities arose in the 1300s and with them notions of academic freedom centered on the book, its reading and analysis. The book has frightened kings and queens so much that they have banned books and their authors. Catherine the Great of Russia considered herself such a woman of the Enlightenment that she corresponded with Voltaire—in French—to share her noble ideas with him; he praised her and published her letters. But she was not so enlightened as to permit Alexander Radishchev, a relatively wealthy nobleman, to go unscathed for publishing his *Journey from St. Petersburg to Moscow* (1790), an attack on privilege and laziness among the Russian elite and on the moral evil of serfdom. She sentenced him to death, although commuted that sentence to exile in Siberia. Books have been burned in fascist society and banned even in that bastion of democracy, the United States, for their alleged pornographic content (Voltaire, Chaucer, Defoe, D. H. Lawrence, Alice Walker, Gabriel García Márquez), but usually for the political discomfort they awaken (Mark Twain's *The Adventures of Huckleberry Finn*, John Steinbeck's *Grapes of Wrath*, J. D. Salinger's *The Catcher in the Rye*, and Toni Morrison's *Beloved*, among scores of others). Imagine my shock to discover that books are under a different kind of attack in the twenty-first century, being considered as not crucial to the modern library because of the electronic age. The central place of books in the history of academic and political freedom requires a better fate than the emptying of stacks for storage or "recycling."

A summer trip once brought me to a French chateau, the home of my friend's uncle, a wealthy but unassuming man, whose summer place had three libraries; "chateau" does not do the place justice. One of the libraries, in the east tower, was for handwritten manuscripts, the other two were libraries for later books. How could you browse in such a place with such smells and textures if using an electronic catalog? Today does any student—or library director—recall the importance of strolling through stacks, browsing, and stumbling upon a treasure of a book?

A wonderful Norwegian film short sits on the Internet. In it, a medieval monk asks another monk (the "help desk") how to read a book. How do you open it? Can you go back to the beginning? Must the spine be on one side to open it? I treasure the moments of holding a book in my hands, looking at the contents, then the index and notes. It's not the same experience as reading a pdf online. Nor does sliding your finger across a screen constitute reading or guarantee comprehension. The pleasure is gone. To put the question another way, would you rather receive a written letter through the post declaring love for you, or an emailed "Luv u"? Do you prefer an online publication that you post yourself or a copyedited book published by a university press and then reviewed—even if it's a nasty review—in some national print magazine or journal?

Electronic publishing is no panacea and it is not inexpensive. Since the 1970s such large publishing conglomerates as Elsevier, Springer, and Wiley have "aggressively" acquired science journals—and charge a great deal for them, "so academic libraries have had to devote considerable financial resources to retain them, and that has diminished their budgets for humanities and social science monographs." The advent of e-books, the decline of bookstores, and other pressures have "put enormous strain on university presses." And universities are no longer subsidizing their presses the way they did at one time,[3] perhaps because there are more administrators than there were twenty years ago. It is foolish and wrong, said Peter Dougherty, director of Princeton University Press, "that scholarship should sort itself out spontaneously over the Internet."[4] We need leadership, not a flippant concession to the Internet as if it is inevitable or good. I shall begin the process straightaway: I shall get a book off my desk, browse through it, dog-ear the pages, and later today let it get wet in the rain. (I shall also enjoy the joys of JSTOR and other online search engines later today, I admit.)

My lament about the assault on scholarship by the Internet is reminiscent of those about other, previous electronic communication technologies. Since the late nineteenth century various social commentators claimed that the telephone would destroy the American family, interfering with time within the nuclear family, drawing children away from proper parental influences, and endangering especially young girls who might be encouraged to engage in courting away from the watchful eye of their parents. Granted, the telephone enables us to be in contact with our loved ones and to get help in emergencies, and in

1909, one observer noted "advantages in urban life which must favorably affect the domestic institution. There are wider and more rapid means of communication and of receiving impressions; although the rural telephone and trolley are making marvelous changes outside the cities. There are more mental stimuli in the thronged street than in the sleepy lanes, and along the quiet waters of pastures and meadows."[5] He may have been wrong about the quiet waters—I found them quite stimulating.

In the 1950s folks worried about similar things, and perhaps with good reason, because of the television. What happened to the family dinner? Why do so many people have three or even four TVs blaring in different rooms? I have another complaint: What happened to in-depth reporting? Television news used to be journalism, but it has become sound bites without any nuance in which "reporters" color any story in black and white terms, as if there are only two sides to a story, or even seek, in the name of "balanced" reporting, to offer crackpots and liars equal time to dispute accounts or facts. The BBC recently instructed its journalists *not* to give equal time to climate change deniers since no reputable scientist working independently of the petroleum-funded Heartland Institute doubts the significant, human, fossil-fuel contribution to rising temperatures, polar melting, rising sea level, and increasingly virulent pandemics. Most viruses and bacteria love warmer temps, have you noticed?

But my real concern is with the "connected generation" and its carefree attitude about the unquestioned goodness and inevitability of electronic information. Because of such devices as the "smart phone," people feel a need also to check their emails, SMS, and apps, even when intimately engaged with another individual. They check their phones when they get up, when at work, on the weekends; they read the same messages over and over, looking for nuance; they archive tens of thousands of them, never to read them again, and certainly not to browse them in the stacks. Twenty percent of young Americans use their cell phone computers while making love; 10 percent of all Americans do so, and 12 percent while in the shower, 35 percent in the movie theater, 33 percent during a date, and 72 percent claim to have their smart phones within two meters of their bodies the majority of the time.[6]

Being connected constantly means people have blurred work and leisure, allowed a handheld device to intermediate their conversations, and forced them to find uses for them when none exist. And they have learned how not to read carefully or think through things fully. I once emailed a student to inform him of his dangerously poor performance in class. He responded nearly instantly, "Dear Prof! I'm sorry. I understand. I'll come see you as soon as I get out of this lecture." Ah, fears confirmed!

These magnificent tools have marvelously enabled contact in emergency or for desire. One need not be alone for a moment. There's always a screen to touch. Yet I sense a direct line between many students' writing problems and the Internet. Without any sense of propriety or knowing how to determine which websites are reputable, having long blurred the distinction between a primary and a secondary source, believing that if it's not one click away on Google, then it doesn't exist, and readily and unthinkingly copying, cutting, and pasting, they plagiarize without realizing it, they adopt the grammatical mistakes of others without noticing, and they have forgotten, like many library directors, whose research careers involve precisely how to organize and accumulate search engines and precisely not how to do research with tactile primary sources, why books and stacks remain crucially important to learning. Like students disconnected from the challenges and joys of writing by cutting and pasting, too many librarians have become disconnected from the craft of research. As for students, if they do not know how to read a book, how to use its scholarly apparatus, how to judge the sources consulted, how to identify, find, and analyze those sources themselves, then how can they properly evaluate the quality, value, and utility of widely available and rich electronic resources? If they do not learn to browse in stacks, how can they open their minds to unexpected and perhaps troubling thoughts? As for librarians, they need to trust the builders of libraries and the users of books to offer advice about how to maintain crucial collections of books and journals in connection with electronic resources so that faculty can teach students how to think, read, and write. Instead, they seek to store books away somewhere and the art of browsing is lost. Some of them also have no conception of how architecture influences the learning process and, succumbing to the impulse to create another shopping mall—the malling of America—they opt for removing books to create open spaces like a food court or huge open waiting room with plugs and outlets to hook into networks of instantaneous yet fleeting gratification. The last two things a student needs in the library to master a subject are another food court or coffee shop and a smart phone. As I anticipated, thankfully, sales of the e-book began to decline in 2014, with "real" books making a rebound. Who wouldn't rather have the tactile as well as intellectual experience that a book provides?

Bicycles

My second example comes from the area of transportation, and in this case the object is a bicycle. I bicycle to work, weather permitting; Maine has lots of ice and snow. One day a week I use my automobile for errands. Otherwise, it sits at home. For many people, there is no alternative to the automobile. Through a series of poor public policy choices, infatuation with the internal combustion engines, and the connivance of automobile manufacturers, the US public transport system is a shell of its former self. From the 1930s until the 1950s, automobile, truck, tire, and oil companies conspired to buy up and dismantle trolley and tram systems in forty-five cities, then convert them to bus operations—from which they directly benefited—and from there to allow the transport systems to decay. As a result very few American cities have adequate public transport. Alone in the world in the belief that public transport should be for-profit operations, American policy makers have worked hard to end support in a misguided effort to cut taxes without thinking of real costs or ways to cut reliance on oil and gas, and with the firm belief that any "business" should profit on its own.

Large-scale public transport dates to the nineteenth century and did not need subsidies. It required subsidies in the second half of the twentieth century when people turned more and more to automobiles which, at first, seemed less expensive than trams, trains, trolleys, and buses. The lower cost at first glance arose from the fact that we rarely consider the cost of automobiles to environmental degradation, pollution, global warming, or such health dangers as lung and heart disease that come from air pollution. Nor have we been honest about the substantial subsidies that automobiles gain in taxes for road, bridge, and other projects, let alone subsidies and write-offs to the highly profitable oil companies. Instead, knee-jerk budget cutters refuse to subsidize public transportation, while the hidden and direct subsidies to the automobile run billions of dollars annually because of a love affair with the automobile. If only they loved bicycles.

As a result, public transportation has disappeared in America; perhaps 2 or 3 percent of all Americans use public transportation on a given day. Even in major metropolitan areas workers commute by personal cars, and usually alone, with the exception of the New York metropolitan area, the Pacific Northwest, San Francisco and Los Angeles, California, and Boston, Massachusetts. In most places there are no options to the private car.[7] Only 865,000 Americans commute by bicycle, less than 1 percent daily of all commuters, although this meager figure was up 9 percent in 2012 over 2011.[8] Americans find it difficult to bicycle in the absence of dedicated bicycle lanes and pathways and lack of understanding of bicycle culture among automobile drivers.

The automobile costs too much in all ways. The Internal Revenue Service calculates the cost of operating a motor vehicle at $0.56/mile based on oil and gas, depreciation, insurance, repairs, and so on. A twenty-mile trip therefore costs $11.20 each way. A forty-mile daily round-trip commute means about $5,000 in annual costs (roughly $100 per week). Imagine the savings of gasoline by using bicycles or public transportation, imagine the cleaner air, the lower road rage. If you commute at $15/day, it's still cheaper than driving your car $100/week. And you could read a big book—or even look at a computer or smart phone screen. Yet the bicycle and the light-rail vehicle find it hard going because of policies and practices geared to enabling the love affair with the violent machine, tax revenues dedicated to those practices, and subsidies to oil.

If roughly half a million automobiles had taken to unpaved roads in the United States in 1910, and there were thirty million smoke-belchers in 1930, then there were over 250 million passenger vehicles in the United States in 2014 that used nearly 4 million miles of roads (2.6 million paved) including 47,000 miles of the magnificent interstate highway system built at a cost over thirty years of $131 billion. I have heard it said that total automobile, truck, and airplane infrastructure in the United States since World War II cost nearly $4 trillion including capital costs and repairs. The United States uses 370 million gallons of gasoline daily.[9] We import much of this oil. We paid $631 billion to the Department of Defense in 2014, much of it to ensure a safe and steady supply of oil from suppliers in the Mideast.

What of the train? Passenger rail by Amtrak involves in all 300 trains daily and a total of 21,300 miles of track (versus 14,000 trains in France . . . or 30,000 trains in Germany . . . both daily). The government provides miserly support to Amtrak, and even the Rail Safety Improvement Act of 2008 that passed despite George W. Bush's veto threat provided only $2.6 billion annually through 2013, and it was then cut nearly in half to $1.4 billion. Amtrak may have received $50 billion over thirty years. Amtrak pays property taxes, receives much less than highways and airways in support—which is the reason for the decline of passenger service, and as a result, of major industrial countries only the United States relies on the automobile. The head of Amtrak estimated that annual subsidies for Amtrak were $40 per passenger, while they were $700 per automobile.[10]

We even subsidize cruise ships directly and indirectly at a total of something like $8 billion annually. The US government helps the Carnival Corporation directly: the Coast Guard "keeps the seas safe for Carnival's cruise ships. Customs officers make it possible for Carnival cruises to travel to other countries. State and local governments have built roads and bridges leading up to the ports where Carnival's ships dock." Between 2006 and 2011 Carnival paid only 1.1 percent of its $11.3 billion in profits for total federal, state, local, and foreign taxes because of loopholes.[11] Or consider the fact that former senators Trent Lott (R-MS) and Daniel Inouye (D-HI) were "prominent supporters" of a Title XI federal loan guarantee program that enabled a billionaire real estate developer to receive a $1.1 billion loan guarantee for two cruise ships to be built in Lott's hometown of Pascagoula, while Inouye offered a provision in a defense bill that gave the company exclusive rights to operate in Hawaii. The company went bankrupt and cost taxpayers $187 million.[12] That's 200,000 bicycles—one for every fifteen Mississippeans or, to be fair, one for every seven Hawaiians.

The ongoing efforts in the House of Representatives to cut funding for public transportation and for Amtrak will exacerbate reliance on oil and lead to the loss of more than 600,000 transit and related jobs. The Highway Trust Fund, some of which goes to mass transit, is losing solvency. The federal gas tax is $0.184/gallon as it has been for decades of inflation, and gas costs twice as much as only a few years ago. To replenish the Fund, to start rebuilding aging bridge and highway infrastructure, to find small support for public transportation in June 2014, two senators proposed a $0.06 increase per gallon twice over two years. But Congress is against any tax and prefers to let the transportation decay as states and local governments are left without funds to do anything. The point is that the train, the trolley, the bus are not merely things-in-themselves, but reflect broader social, political, and cultural concerns and goals. Unfortunately, few Americans seem to share them. And they seem positively fearful of bicycles.

From the point of view of the bicycle, it would be lovely to live in the Netherlands. I understand the Netherlands is geographically small, but Dutch people have no qualms about biking along safe and well-maintained bikeways for fifteen kilometers or more to work or school. It's a healthy endeavor. The Netherlands is the "cycling nation" where bike trips surpass automobile trips, where 26 percent of the people cycle.[13]

Not even the annual World Naked Bike Ride (June 14 in a city near you if you haven't noticed) has generated interest in cycling among American drivers.

How can I so love the bicycle given its genesis? In many ways, it begat the automobile. Through machine tools for grinding and gear cutting and gears, through innovations in bearings, through light and less expensive steel production, steel tubes, pneumatic tires, chain drives, and differential gearing, it was a technological precursor of the automobile.[14] It was a crucial and important step to mass production because bicycle shops were among the first to recognize the need for specialized machinery and interchangeable parts, punch pressing and stamping instead of (exclusively) welding.[15]

Instead of helping the bicycle and the pedestrian, we have spent years stamping public places with the automobile, turning the street from a market- and meeting-place into a thoroughfare, building roadways without nearby sidewalks, making it impossible for people to bike safely, let alone walk to supermarkets or schools.[16] In many of his essays, Lewis Mumford decried the architectural desiderata of the automobile for American cities.[17]

Some regulations and technologies exist to "calm" traffic, decrease the numbers of individuals on the roads, make it more costly for them to drive, and, perhaps as a last resort, lead them down the terrible, dark parkway to public transportation. Rather than make people pay the true costs of driving to work every day, driving to the store every day, heavens—driving to the driving range—state officials have sought such ineffective solutions as the "high occupancy vehicle lane" (HOV)—lanes dedicated to vehicles carrying multiple occupants. Of course, at first they specified four occupants, then three, and now two. But even the two-occupant requirement has very few takers or makes much of a difference. It is fairly clear that time savings from HOVs are insufficient or that many carpoolers live and/or work together. HOV lanes date to the early 1970s in the United States and were first employed in the Washington, DC, area; supporters claim that they can save 50 percent of travel time into the nation's capital during rush hour over conventional lanes.[18]

In the early 1970s, I worked at the Center for Auto Safety, an organization created in part with funds Ralph Nader had won in a suit against GM for illegally surveilling him and trying to entrap him with prostitutes after he published *Unsafe at Any Speed* (1965) in which he described the lags in auto safety in America, including for GM's ugly, poorly engineered, rear-engined monster, the Corvair. I remember reading a series of studies in the Center library that proved that road improvements to ease traffic through widening roads were doomed to failure since they inevitably attracted more drivers who saw the improvement as a panacea, filled the new road with more vehicles, and created the same problem again. Some people call this, after the film *Field of Dreams*, the "if-you-build-it-they-will-come" syndrome. Former Bogotá, Colombia, Mayor Enrique Penalosa said that trying to solve traffic problems by building roads is like trying to put out a fire with gasoline.[19] In spite of the beauty of the $25 billion traffic improvement, Boston's legendary "Big Dig"–induced traffic has led to no net improvement for Boston vehicle commuters.[20] Imagine if that $25 billion had been spent on parks, greenways, public transport, and bicycle paths.

Perhaps responding in part to the statement of the one-time supporter of universal health care as governor of Massachusetts, dog lover, and former presidential candidate Mitt Romney, that "corporations are people," one such HOV pioneer, Jonathan Frieman, fought a $478 traffic ticket for driving alone in a California Highway 101 HOV lane because, he pointed out, his "passenger was the articles of incorporation of his business—which were placed unbuckled on the driver's seat." But Frieman may have a point—even if he will destroy the HOV attempt to ease traffic, since in 2010 the US Supreme Court, he noted, treated corporations as people by granting them First Amendment rights to contribute freely to political campaigns. Frieman also points out that the California state vehicle code defines a person as "natural persons and corporations." Thankfully, the traffic court hearing officer, Frank Drago, "wasn't sympathetic." He ruled that "common sense says carrying a sheath of papers in the front seat does not relieve traffic congestion. And so I'm finding you guilty."[21] I do love, however, EZPass, electronic eye toll-paying devices, for cutting lines at toll booths and thereby lowering gas consumption.

Road improvements, in other words, induce more traffic. According to the Campaign for Sensible Transportation, HOVs have not brought more buses to the roads. In fact, they bring fewer fare-paying commuters, leading to reduced fares and services, therefore more people abandoning public transportation, and ultimately to even more congestion.[22] The only solution is traffic calming, efforts to slow the automobile further, the determination to shed lanes, narrow and squeeze traffic—and build dedicated bicycle lanes.

In summary, I propose that 3–5 percent of every "road improvement" budget or new construction budget be applied to building a dedicated pedestrian-cyclist pathway next to the road, safely protected by guardrails, well-marked, as a symbol of hope for America's future and for the benefit of cyclists everywhere, no matter their political views, and with trees and wild flowers planted along the resulting greenway, with flowers also to be chosen irrespective of politics, although perennials like black-eyed susans would be nice. I also oppose efforts to straighten any road or add lanes to it. I embrace the passenger train for its romance and its clickety-clack sound. And I recommend highly the speed bump to all city planners interested in the future of their communities. But this is the subject of my next book.

Five big lessons therefore come to mind from this book. First, follow the money to understand the extent to which vested interests put those innocuous objects on your breakfast table, in your refrigerator, at your fingertips, in your garage, and on your body. Second, recognize that all of them are international to one degree or another. They are traded internationally, they find their way, almost autonomously, into food supplies, and they often involve unsafe, unhealthful, and environmentally unsound production and labor practices abroad that are outlawed at home. Third, learn how to live without many of these objects, or at least put them aside from time to time, no matter how advertisers and industrial psychologists try to convince us of their indispensability. Fourth, and paradoxically, many technologies can free us from the inconvenience and inequity that others impose. I think of the bicycle for common sense and the speed bump for traffic calming. Fifth, sometimes you should just say "no," refuse that new-fangled fish stick or aluminum soda can or smart phone or online source. Buy a baguette, grab a book and a blanket, get on your bicycle, go to a park, and wrap yourself around both of them—the book and the blanket.

Notes

INTRODUCTION: Technostories

1. Thomas Hughes, "The Evolution of Large Technological Systems," in *The Social Construction of Technological Systems*, ed. Wiebe Bijker, Thomas Hughes, and Trevor Pinch (Cambridge and London: MIT Press, 2012), pp. 45–76. See also Hughes, *Rescuing Prometheus* (New York: Pantheon, 1998).

2. David Nye, *Consuming Power* (Cambridge, MA: MIT Press, 1998), pp. 2–3.

3. Alvin Weinberg, the former director of Oak Ridge National Laboratory, and a promoter of innovations in power reactor design, considered, among other factors, "momentum" in the organization and financing of big science and technology in his *Reflections on Big Science* (Cambridge, MA: MIT Press, 1968).

CHAPTER 1: The Ocean's Hot Dog

Acknowledgments. I would like to thank E. Robert Kinney, former president of General Foods; Professor Marcus Karol of the Massachusetts Institute of Technology's food science program; Professor Deborah Fitzgerald of MIT's STS program; and Professor Raffael Scheck of Colby College for their comments on this chapter; Kaitlin McCafferty and Carrie Ngo for their research assistance; and the two anonymous reviewers for their patient and careful suggestions on the version of this chapter originally published in *Technology and Culture* in 2008.

Epigraph. Ludwig Feuerbach, "Der Mensch ist, was er isst," quoted in Hans Werner Wüst, *Das Grosse Zitaten Lexikon* (Vienna, 2004), 81.

1. On the industrialization of the postwar food industry, see Harvey Levenstein, *Paradox of Plenty: A Social History of Eating in Modern America* (New York: Oxford University Press, 1993), 101–18.

2. Birds Eye is now owned by Unilever and sold under the Iglo, Birds Eye, and Findus brands; Gorton's has been a subsidiary of Nippon Suisan Kaisha, Ltd., since 2001. "Dinner, Frozen or Dried," *Newsweek*, Nov. 19, 1945, 72–74; Don Wharton, "Birdseye Also Means Man," *Reader's Digest*, Dec. 1946, 71–74; and Levenstein, *Paradox of Plenty*, 106–7.

3. By 1926, Clarence Birdseye had established a twenty-ton quick-freeze operation in Gloucester that produced a crude form of the fish block, which was later to become the "ore" of the fish stick industry.

4. "Drip Control in Frozen Fish," *Food Industries* 13 (1941): 100; J. Perry Lane, "Time-Temperature Tolerance of Frozen Seafoods," *Food Technology* 18 (1966): 156–62; and "The

'Multiplier Effect' of Frozen Food Technology on American Life," *Food Technology* 15 (1961): 14–24.

5. Louis Berube, "Modern Practice of Fish by Cold," *Food Industries* 9 (1937): 645; J. C. Bauernfeind, E. G. Smith, and G. F. Siemers, "Commercial Processing of Frozen Fish with Ascorbic Acid," *Food Technology* 5 (1951): 254–60; "Drip Control in Frozen Fish"; and D. D. Gillespie, J. W. Boyd, H. M. Bissett, and H. L. A. Tarr, "Ices Containing Chlorotetracycline in Experimental Fish Preservation," *Food Technology* 9 (1955): 296–300.

6. F. W. Knowles, "How Foods Are Frozen in the Northwest," *Food Industries* 12 (1940): 54–56, and "Reduces Drying in Air Blast Freezing," *Food Industries* 13 (1941): 92–93.

7. Gerald Fitzgerald, "Why You Freeze It That Way," *Food Industries* 22 (1950): 73–77. Another way to limit drip had nothing to do with additives. Chemists at the Bureau of Commercial Fisheries determined that the excess cutting of fish caused millions of cells to rupture, thus exposing their contents to the atmosphere. See Frederick King, "Cell Damage from Excess Cutting of Fish Adversely Affects Frozen Seafood Quality," *Quick Frozen Foods* (1962): 115–16.

8. Ivan Miller, "Quick Freezing Thaws Frozen Channels of Distribution," *Food Industries* 10 (1938): 199, 202.

9. "Fishbricks for Fastidious Housewives," *Business Week*, July 12, 1947, 64.

10. Francis Schuler, "The Papal Decree, Kennedy Round Present Fish Sticks, Portions with Challenge," *Quick Frozen Foods* (1968): 152, 188–89.

11. "Dinner, Frozen or Dried," 72–74, and Wharton, "Birdseye Also Means Man," 71–74.

12. Miriam Wright, *A Fishery for Modern Times* (Toronto, Canada: University of Toronto Press, 2001).

13. "Two New Freezer Factory Ships May Put U.S. Back into the Fish Industry Race," *Quick Frozen Foods* (Sep. 1968): 97–98. On the economic, environmental, and social costs of the modern trawling industry in the North Atlantic, see Michael Harris, *Lament for an Ocean* (Toronto, Canada: McClelland & Stewart, Inc., 1998); Mark Kurlansky, *Cod: A Biography of the Fish That Changed the World* (New York: Walker and Company, 1997); William Warner, *Distant Water: The Fate of the North Atlantic Fisherman* (Boston, MA: Penguin Books, 1983); and David Dobbs, *The Great Gulf: Fishermen, Scientists, and the Struggle to Revive the World's Greatest Fishery* (Washington, DC: Island Press, 2000).

14. E. Robert Kinney, personal communication, Oct. 20, 2003; see also "Better Seafood Possible with New Fishing Boat," *Science News Letter*, Dec. 5, 1953, 359.

15. Garrett Harden, "The Tragedy of the Commons," *Science* 162 (1968): 1243–48.

16. "Progress in Freezer Rail-Transportation," *Food Industries* 10 (1938): 62.

17. "Fresh and Frozen on the Same Delivery," *Food Industries* 17 (1945): 86–87.

18. "The 'Multiplier Effect,'" 14–24.

19. Lane, "Time-Temperature Tolerance," 197–201.

20. Max Zimmerman, *The Super Market* (New York: McGraw Hill, 1955); Ralph Cassady, *Competition and Price Making in Food Retailing* (New York: Ronald Press, 1962); and Edward Brand, *Modern Supermarket Operation* (New York: Fairchild Publications, 1963).

21. Levenstein, *Paradox of Plenty*, 113–14. The supermarket held a central place not only at home, but also in the ideology of Cold War competition. The supermarket display case figured prominently in US exhibitions at international trade fairs, as it did in the public sparring between Vice President Richard Nixon and Soviet Premier Nikita

Khrushchev in a debate in an American kitchen at a Moscow exhibition in July 1959. See also Ruth Oldenziel and Karin Zachmann, eds., *Cold War Kitchen: Americanization, Technology, and European Users* (Cambridge, MA: MIT Press, 2009).

22. "Food Fair's Seafood Distribution Center Stresses Frozen," *Quick Frozen Foods* (1960): 86, 107.

23. Levenstein, *Paradox of Plenty*, 101–6.

24. Schuler, "The Papal Decree," 152, 188–89.

25. Laura Shapiro, *Something from the Oven* (New York: Penguin Group, 2004), 8–22, 24–27.

26. "New Product Seen Spurring Fish Use," *New York Times*, Oct. 3, 1953.

27. "E. Robert Kinney, Former General Mills CEO '39 Repaying 'Debt' to Bates," http://www.bates.edu/alumni-kinney.xml (accessed Aug. 20, 2007).

28. G. E. Petley to Paul M. Jacobs, Gorton-Pew Fisheries Co., Ltd., Oct. 9, 1953, Gorton's Corporate Archives (hereafter GCA).

29. Welles Sellew, memorandum, 3 July 1953, GCA (emphasis in original).

30. Director of Sales Promotion Paul Jacobs, memorandum, Dec. 1, 1953, GCA.

31. Paul Jacobs, "The Fabulous Fish Stick Story," typescript (c. 1952), GCA.

32. Ibid.

33. Fish stick brochure (c. 1954), GCA; Robert Vanderkay, "The Fine Edge in Continuous Cutting of Frozen Seafoods for Processing," *Quick Frozen Foods* (Nov. 1966): 74–76; and Berube, "Modern Practice," 645. Most engineers argued that mechanization enabled them to fully control workers and the production process.

34. Lane, "Time-Temperature Tolerance," 197–201.

35. GCA, various undated Gorton's salesmen brochures; Welles Sellew to all brokers, memorandum, July 3, 1953; and Paul M. Jacobs to all brokers, memorandum, Dec. 1, 1953.

36. Lee Geist, "Fish Fillip," *Wall Street Journal*, May 7, 1954, 1, 11.

37. "Fish Stick Expansion," *The Man at the Wheel*, Dec. 1954, 1.

38. "Fish Sticks: A Survey Made among LHJ Subscribers," *Ladies' Home Journal Research Department*, Nov. 1954, in GCA.

39. Internal memorandum, "Who Buys the Food and Who Prepares It? Women" (c. 1958), GCA.

40. *The Man at the Wheel*, March 1960, 2.

41. "Frozen Fish Grading Offered in Bill," *Food Industries* 22 (1950): 147.

42. "Frozen Fish Sticks," *Consumer Reports*, March 1956, 63–65; and "Frozen Fried Fish Sticks," *Consumer Reports*, Feb. 1961, 80–83.

43. Francis Schuler, "History and Future of Fillet-Based Frozen Fish," *Quick Frozen Foods* (May 1968): 94–95.

44. J. J. Powell, "Pre-Standard Educational Program," *Quick Frozen Foods* (April 1960): 130.

45. J. R. Brooker, "Six Years of Frozen Seafood Inspection," *Quick Frozen Foods* (Aug. 1964): 131–32.

46. "Frozen Fish Stick Quality Blasted in February 1961 Consumer Reports," *Quick Frozen Foods*, April 1961, 136.

47. J. R. Brooker, "USDA Frozen Seafood Inspection," *Quick Frozen Foods* (Sep. 1967): 99, 167.

48. "Should Mandatory Frozen Seafood Inspection Replace Voluntary Inspection?" *Quick Frozen Foods* (Jan. 1966): 95–98 (emphasis in original).

49. "Number of Frozen Seafood Processors Increased 774% Over Past 20 Years," *Quick Frozen Foods* (Nov. 1967): 265–66.

50. E. Robert Kinney, "Government Inspection Vital Facet of Search for Improved Products," *Quick Frozen Foods* (May 1968): 95–96.

51. "Your Company," *The Man at the Wheel*, March 1954. To this day, the National Marine Fisheries Service supports research, development, and marketing strategies for fisheries and fishery products under the act through "grants for research and development projects addressed to any aspect of US commercial and recreational fisheries including, but not limited to, harvesting, processing, marketing, and associated infrastructures" (Grants Office, National Marine Fisheries Service, http://www.ago.noaa.gov/grants/ [accessed Aug. 2, 2006]). Funding for the program was severely reduced by the Bush administration in 2002.

52. Gordon Gunderson, "History of the National School Lunch Program," in *National School Lunch Program Background and Development, Food and Nutrition Service* (FNS) 63 (Washington, DC, c. 1969).

53. Kathryne I. Sheehan, "Trends in the School Lunch Program," *Journal of Home Economics* 44 (1952): 697–700.

54. "Fish Sticks 'n' Chips," Gorton's pamphlet, mid-1950s, GCA.

55. Stan Hurley, memoranda to brokers, Aug. 16 and Sep. 24, 1954. The cover of a Gorton's brochure from 1954 featured a young girl in a Newton, Massachusetts, school, intended to portray someone typical of the "10 million children who eat school lunches every day. She is ready to enjoy what is rapidly becoming a favorite lunch of school children . . . Gorton's Cooked Fish Sticks." See "A Smile," *The Man at the Wheel*, Nov. 1954. On school lunches, see also Grace E. Hochmuth and Bessie West, "Organizing a School Lunch Program," *Journal of Home Economics* 41 (1949): 559–60; George Hecht, "Save the School Lunch Program!" *Parents* 23 (1948): 14; Levenstein, *Paradox of Plenty*, 78; Sheehan, 697–700; and http://www.scusd.edu/nutrition_education/history_of_school_ lunch_program.htm (accessed Aug. 20, 2007).

56. Harvard University report, typescript (1954), 27–33, 36–40, GCA.

57. Levenstein characterizes the postwar years as the "golden age" of food chemistry, when scientists developed hundreds of additives, preservatives, color stabilizers and dyes, and "smootheners." Fish sticks had the built-in advantage among processed foods that cod and other fish used in them could not be farmed, but only industrially harvested from the ocean (Levenstein, *Paradox of Plenty*, 109–12).

58. Paul Jacobs, memorandum (c. 1956), GCA: "I am a fish peddler, and as such I am not invited to illustrious gatherings such as this. There has been a stigma against my industry. . . . I realize it all-to-well in my own family life. My daughter came home from school recently looking somewhat crestfallen. . . . She told me the teacher had asked all the children what their fathers' occupations were. When my daughter, Susie, said her father sold fish, all the children said . . . 'Gee, he must stink!' So, to solve a family problem I changed my vocation. Gentlemen, I stand before you as a piscatorial engineer!"

59. "Frozen Foods Seminar Held," *Boston Herald*, July 20, 1956; and Gorton's History, 133–34.

60. Samuel Goldblith, "Introduction," in *Exploration in Future Food Processing Techniques*, ed. Samuel Goldblith (Cambridge, MA: MIT Press, 1963), 1–2. The National Canners' Association Laboratory was established in 1913 under the direction of E. J. Cameron, an MIT graduate.

61. Stephen Lirot, "A Study of Some Factors Affecting the Quality of Precooked Fish Sticks" (master's thesis, MIT, 1955); and Robert Goldthwaite, "A Study of the Oxidation of Fat in Fried Frozen Fish Sticks" (bachelor's thesis, MIT, 1961).

62. "Gorton's Unveils Its 'Fresh-Lock' Process," *Quick Frozen Foods* (Sep. 1962): 108–9. "Fresh-lock" was a patented dip process by which natural chemicals forming a gel-like solution of protein interacted with the cells of the fish and eliminated drip loss.

63. William Phelps, "MIT Scientists Preserve Food by One-Second Shots of Cathode Rays," *Boston Sunday Post*, May 24, 1953; "Radiation-Pasteurization Extends Storage of Fresh Seafoods Significantly Longer," *Quick Frozen Foods* (April 1966): 126–30, 196; John Nickerson, "Radiation-Pasteurization of Marine Products," in *Exploration in Future Food Processing Techniques*, 46–53; and "Irradiation Preservation of Pacific Northwest Fish," *Food Technology* 14 (1960): 411.

64. Dave Rhinelander, "Seafood Center's Birthday: Seminar Speakers Arrive," *Gloucester Daily Times*, Aug. 25, 1961, 1, 14, and "Management, Labor, Government Chart Seafood Industry Growth," *Gloucester Daily Times*, Aug. 26, 1961, 1, 10.

65. Rhinelander, "Seafood Center's Birthday."

66. "Number of Frozen Seafood Processors Increased 774% Over Past 20 Years," 265–66.

67. "Portion Production Hits New High," *Quick Frozen Foods* (March 1963): 182.

68. "Frozen Fish Near 50% of Total Consumption," *Quick Frozen Foods* (July 1965): 129–30; *Quick Frozen Foods* (January 1966): 101; *Quick Frozen Foods* (April 1966): 133–34; "U.S. Seafood Per Capita Consumption Up in 1968," *Quick Frozen Foods* (April 1969): 139–40; "1967 Fish Portion Production Up," *Quick Frozen Foods* (May 1968): 93; and "Frozen Seafood Reaches Record High," *Quick Frozen Foods* (June 1970): 79–81, 85, 120–21.

CHAPTER 2: The Sports Bra

Acknowledgments. Many of the manufacturers of sports bras understandably have been hesitant to share extensive information about the science, technology, and design of their product because of the competitive nature of the market. To a certain extent, therefore, the sources for this history are limited. I am thankful to representatives of several companies for their willingness to talk history with me. I would also like to thank Jen Goebel, Cheryl Hile, Julie Millard, Linda Soresi, and Sonja Thomas for their comments on an early draft of this chapter, and Hinda Miller and LaJean Lawson for sharing their great knowledge of the history of the sports bra. I have run more than one hundred marathons and have seen tens of thousands of female runners leave me in the dust.

1. Hinda Miller, *Pearls of a Sultana* (Randolph, VT; Caspar, CA: Public Press, 2012), p. 42.

2. Ibid., pp. 43–44.

3. On the history of the brassiere, see Jane Farrell-Beck and Colleen Gau, *Uplift: The Bra in America* (Philadelphia: University of Pennsylvania Press, 2002).

4. Miller, *Pearls*, pp. 45–46. Polly Smith left the business to pursue design, eventually working on costumes for Jim Henson's Muppets, while Hinda bought out her shares and Linda and Hinda became equal owners. See ibid., p. 49.

5. Interview with Hinda Miller, Jan. 7, 2013, by telephone.

6. Interview with Dr. LaJean Lawson, Jan. 11, 2013, by telephone.

7. For two among many fine studies that consider various aspects of science, technology, and gender, see Londa Schiebinger, *The Mind Has No Sex? Women and the Origins of Modern Science* (Cambridge, MA: Harvard University Press, 1989), and Cynthia Daniels, *Exposing Men: The Science and Politics of Male Reproduction* (New York: Oxford University Press, 2006). For a comprehensive retrospective essay on technology and gender, see

among her many articles and books Ruth Schwartz Cowan, "From Virginia Dare to Virginia Slims: Women and Technology in American Life," *Technology and Culture* 20, no. 1 (Jan. 1979): 51–63.

8. For a collection of essays on technology and gender, see Nina Lerman, Ruth Oldenziel, and Arwen Mohun, eds., *Gender and Technology: A Reader* (Baltimore: Johns Hopkins University Press, 2003). A vast new literature on gender, engineering, computers, and gendered spaces has appeared since the 1990s.

9. Andrea Tone, *Device and Desires: A History of Contraception in the United States* (New York: Hill and Wang, 2001).

10. Ruth Schwartz Cowan, *More Work for Mother: The Ironies of Household Technology from the Open Hearth to the Microwave* (New York: Basic Books, 1983).

11. Arwen P. Mohun, *Steam Laundries: Gender, Work, and Technology in the United States and Great Britain, 1880–1940* (Baltimore: Johns Hopkins University Press, 1999).

12. For a discussion of the extent to which the American kitchen found a home in Europe, the evolution of kitchen design in Europe, how the kitchen in the late 1950s was the focus of a dispute between the USSR and the United States over the advantages and costs of consumer society, and other important questions, see Ruth Oldenziel and Karin Zachmann, eds., *Cold War Kitchen: Americanization, Technology, and European Users* (Cambridge, MA: MIT Press, 2009).

13. Schiebinger, *The Mind Has No Sex?*

14. Stephen Gould, *The Mismeasure of Man* (New York: W. W. Norton, 1981).

15. Donna Haraway *Primate Visions* (New York and London: Routledge, 1989).

16. *Women, Minorities, and Persons with Disabilities in Science and Engineering*, published by the National Science Foundation, roughly every other year since 1994.

17. Interview with Dr. LaJean Lawson, Jan. 11, 2013, by telephone.

18. Kelly Bastone, "A History of the Sports Bra," *Women's Adventure Magazine*, March/April 2007, http://www.womensadventuremagazine.com/issues/the-sports-bra-turns-30/?print=1.

19. Ibid.

20. On the evolution of the business, see Miller, *Pearls*, pp. 51–61.

21. On this subject see, for example, Judith Lorbert, "Believing Is Seeing: Biology as Ideology," *Gender and Society* 7, no. 4 (Dec. 1993): 568–81.

22. Bastone, "A History of the Sports Bra."

23. Interview with Hinda Miller, Jan. 7, 2013, by telephone.

24. Ibid.

25. Ibid.

26. Ibid.

27. Paul Heintz, "So Long, 'Sultan': Sen. Hinda Miller Stages Her Exit," May 20, 2012, http://www.7dvt.com/2012so-long-sultana-sen-hinda-miller-stages-her-exit.

28. Miller, *Pearls*, p. 40.

29. Email from Jill Zanger, Marketing, Nike, Aug.10, 2011. Jill worked with Dr. Susan Sokolowski, senior manager of Apparel Innovation for Nike, who answered questions on the technical innovation background and information.

30. "Nike Offers High-Tech Inner Actives Sports Bra," March 22, 1999, http://adage.com/article/news/nike-offers-high-tech-actives-sports-bra/24878/. According to Nike, "Functional needs and the rise of women participating in a variety of sports and fitness activities drove the development of our first sports bra."

31. Personal communication with the author, email, Jan. 8, 2013.

32. The site "Sweaty Betty" helps women select among various models. See "Sweaty Betty, Bra Guide," http://www.sweatybetty.com/Help/Help.asp?page=braguide&title =bra+guide. Click on cup size and nature of activity for recommendations.

33. Email from Jill Zanger, Marketing, Nike, Aug. 10, 2011.

34. Interview with Dr. LaJean Lawson, Jan. 11, 2013, by telephone.

35. American Kinesiology Association, "Kinesiology Journals," http://www.american kinesiology.org/kinesiology-journals.

36. Lawrence Brawley et al., "Sports Psychology Sex Bias in Evaluating Motor Performance," *Journal of Sport and Exercise* 1, no. 1 (March 1979): 15–24.

37. Sir Astley Paston Cooper first noted the role of this ligament in *On the Anatomy of the Breast* (London: Longman, Orme, Green, Brown and Longmans, 1840).

38. K-A. Page and J. R. Steele, "Breast Motion and Sports Brassiere Design: Implications for Future Research," *Sports Medicine* 27, no. 4 (April 1999): 205–11.

39. Janet Cromley, "Engineered to Beat the Bounce," May 8, 2006, http://articles .latimes.com/2006/may/08/health/he-bras8.

40. Shock Absorber, "Why a Sports Bra?," http://www.shockabsorberusa.com/why.php.

41. Bastone, "A History of the Sports Bra."

42. American Medical Women's Association, "Christine E. Haycock, MD," https:// www.amwa-doc.org/faces/christine-e-haycock-md/ accessed Sep. 23, 2014.

43. Bastone, "A History of the Sports Bra."

44. Ibid.

45. Interview with Dr. LaJean Lawson, Jan. 11, 2013, by telephone.

46. D. Lorentzen and L. Lawson, "Selected Sports Bras: A Biomechanical Analysis of Breast Motion While Jogging," *Physician and Sports Medicine* 15 (1987): 128–39.

47. Interview with Dr. LaJean Lawson, Jan. 11, 2013, by telephone.

48. Cromley, "Engineered to Beat the Bounce."

49. LaJean Lawson and Deana Lorentzen, "Selected Sports Bras: Comparisons of Comfort and Support," *Clothing and Textiles Research Journal* 8 (1990): 55–60.

50. Lara Marks, *Sexual Chemistry: A History of the Contraceptive Pill* (New Haven: Yale University Press, 2001), and Tone, *Devices and Desires.*

51. Institute of Medicine, *Women and Health Research: Ethical and Legal Issues of Including Women in Clinical Studies*, vol. 2 (Washington, DC: National Academy of Sciences Press, 1999).

52. Bruce Mason, Kelly-Anne Page, and K. Fallon, "An Analysis of Movement and Discomfort of the Female Breast During Exercise and the Effects of Breast Support in Three Cases," *Journal of Science and Medicine in Sport* 2, no. 2 (June 1999): 134–44.

53. D. E. McGhee et al., "Bra/Breast Forces Generated in Women with Large Breasts While Standing and During Treadmill Running: Implications for Sports Bra Design," *Applied Ergonomics* 44 (2012): 112–18.

54. Anne Casselman, "The Physics of Bras," Nov. 2005, *Discover*, http://discovermagazine .com/2005/nov/physics-of-bras#.UOYFBORDLmc.

55. Ibid.

56. "Bouncing Breasts Spark New Bra Challenge," Sep. 23, 2007, http://www.science daily.com/releases/2007/09/070915124901.htm.

57. Jenny White, Joanna Scurr, and Wendy Hedger, "A Comparison of Three-dimensional Breast Displacement and Breast Comfort During Overground and Treadmill Running," *Journal of Applied Biomechanics* 27, no. 1 (2011): 47–53. See also K-A. Bowles, J. Steele, and R. Chaunchaiyakul, "Do Current Sports Brassiere Designs Impede

Respiratory Function?" *Medicine and Science in Sports and Exercise* 37, no. 9 (Sep. 2005): 1633–40.

58. L. Wood et al., "Predictors of Three-dimensional Breast Kinematics During Bare-breasted Running," *Medicine and Science in Sports and Exercise* 44, no. 7 (July 2012): 1351–57. See also Jenny White and Joanna Scurr, "Evaluation of Professional Bra Fitting Criteria for Bra Selection and Fitting in the UK," *Ergonomics* 55, no. 6 (2012): 704–11.

59. In one chapter of *Maneuvers*, Cynthia Enloe discusses how the US military struggles with clothing for women soldiers, for example where to put a breast pocket and where should skirt length fall. See Enloe, *Maneuvers* (Berkeley: University of California Press, 2000).

60. "Bouncing Breasts."

61. Deirdre McGhee and Julie Steele, "Breast Elevation and Compression Decreases Exercise-Induced Breast Discomfort," *Medicine and Science in Sports and Exercise* 42, no. 7 (2010): 1333–38.

62. Deirdre McGhee and Julie Steele, "Optimising Breast Support in Female Patients through Correct Bra Fit," *Journal of Science and Medicine in Sport* 13, no. 6 (2010): 568–72. See also Kelly-Ann Bowles, Julie R. Steele, and Bridget J. Munro, "Features of Sports Bras that Deter Their Use by Australian Women," *Journal of Science and Medicine in Sport* 15, no. 3 (May 2012): 195–200.

63. Deirdre McGhee, Julie Steele, and Bridget Munro, "Education Improves Bra Knowledge and Fit, and Level of Breast Support in Adolescent Female Athletes," *Journal of Physiotherapy* 56, no. 1 (2010): 19–24.

64. Deborah Franklin, "Busted! Let's Uplift the Truth and Separate the Myth from All Those Reasons Mother Gave Us for Wearing a Bra," May 26, 1993, http://articles .chicagotribune.com/1993-05-26/entertainment/9305260228_1_breast-size-mammary -bra/2.

65. Bastone, "A History of the Sports Bra."

66. Enell Sports Bras, http://www.enell.com.

67. Personal communication with the author, email, Jan. 6, 2013.

68. Anne Casselman, "The Physics of Bras."

69. Bastone, "A History of the Sports Bra."

70. Jen M. L., "Sports Bras: An Open Letter to Their Manufacturers," Oct. 29, 2012, http://www.huffingtonpost.com/jen-ml/sports-bras_b_1964156.html.

71. "Shock Absorber," The Shock Absorber Sports Institute," http://www.shock-absorberusa.com/science.php.

72. Ibid.

73. D. McGhee, B. Power, and J. Steele, "Does Deep Water Running Reduce Exercise Induced Breast Discomfort?" *British Journal of Sports Medicine* 41, no. 12 (2007): 879–83.

74. Personal communication with the author, email, Jan. 7, 2013.

75. Email from Jill Zanger, Marketing, Nike, Aug. 10, 2011..

76. Ibid.

77. Kelly Bastone, "A History of the Sports Bra."

78. Interview with Dr. LaJean Lawson, Jan. 11, 2013, by telephone.

79. Interview with Audrey Kirkland, New Balance, Sports Bra Manager, by telephone, Aug. 9, 2011.

80. Personal communication with the author, email, Jan. 10, 2013.

81. Personal communication with the author, email, Jan. 6, 2013.

82. Interview with Hinda Miller, Jan. 7, 2013, by telephone.

83. Hinda Miller, "Jogbra and Beyond," http://www.msmoney.com/mm/success _stories/jogbra_beyond.htm.

84. "Title IX and Sex Discrimination," http://www2.ed.gov/about/offices/list/ocr /docs/tix_dis.html.

85. Instead, the focus is on the necessity for women to have equal opportunities as men on the whole, not on an individual basis. With regard to intercollegiate athletics, there are three primary areas that determine whether an institution is in compliance: athletic financial assistance; accommodation of athletic interests and abilities; and other program areas, where appraisal of compliance is on a program-wide basis, not on a sport-by-sport basis.

86. "Department of Education repeals Bush-era policy on Title IX," http://usatoday30 .usatoday.com/sports/college/2010-04-20-title-ix-policy-repealed_N.htm?csp=34sports.

87. NCAA, "Title IX Legacy Goes Beyond Numbers," June 22, 2012, http://www .ncaa.com/news/ncaa/article/2012-06-21/title-ix-legacy-goes-beyond-numbers. On top of this, in 1994, women received 38 percent of medical degrees, compared with 9 percent in 1972; women earned 43 percent of law degrees, compared with 7 percent in 1972; and women earned 44 percent of all doctoral degrees to US citizens, up from 25 percent in 1977.

88. *Automobile Workers v. Johnson Controls, Inc.*, http://www.oyez.org/cases/1990-1999 /1990/1990_89_1215.

89. *Ledbetter v. Goodyear Tire and Rubber Company*, http://www.oyez.org/cases/2000 -2009/2006/2006_05_1074.

90. Bastone, "A History of the Sports Bra."

91. Title Nine, "Bras and Undies," http://www.titlenine.com/category/sports-bras -and-undies/high-impact-sports-bras.do?nType=2.

92. One of many recent sports bra patent applications: "Athletic Bra with Adjustable Support System," 7 pages, filing date Sep. 7, 1999, http://www.google.com/patents?hl=en &lr=&vid=USPAT7435155&id=TyuvAAAAEBAJ&oi=fnd&dq=Dri-FIT&printsec =abstract#v=onepage&q&f=false.

93. According to anecdotal evidence and to more comprehensive tests, the Fiona sports bra "is extremely comfortable and supportive, without feeling too bulky. Another advantage is that it is a back-closure bra." "Moving Comfort Fiona Sports Bra," May 16, 2014, at http://running.about.com/od/sportsbrasforrunning/tp/sportsbraforlargechests .htm.

94. Interview with Hinda Miller, Jan. 7, 2013, by telephone.

CHAPTER 3: Sugar, Bananas, and Aluminum Cans

Epigraph. Carl Sandburg, "Old-Fashioned Requited Love," *Poetry* 13, no. 1 (Oct. 1918): 23.

1. Theodore Sealy and Herbert Hart, *Jamaica's Banana Industry: A History of the Banana Industry with Particular Reference to the Part Played by the Jamaica Banana Producers Association, Ltd.* (Kingston: Jamaica Banana Producers, 1984), pp. 102–6.

2. Horace Campbell, "Jamaica: The Myth of Economic Development and Racial Tranquility," *Black Scholar* 4, no. 5 (Feb. 1973): 16–23.

3. Ibid.

4. Joseph E. Inikori, "The Atlantic World Slave Economy and the Development Process in England, 1650–1850," http://www.waado.org/nigerdelta/documents/slavery/slavery anddevelopment-inikori.html.

5. A. J. R. Russell-Wood, "Technology and Society: The Impact of Gold Mining on the Institution of Slavery in Portuguese America," *Journal of Economic History* 37, no. 1 (March 1977): 59–67.

6. Ibid., pp. 73–74, 76.

7. After Eli Whitney invented the cotton gin that inexpensively picked seeds out of cotton before milling, many plantation owners added to or expanded their cotton fields and bought more slaves to pick it. The number of slaves grew by hundreds of thousands. Walton points out that the northward migration of blacks during and after World War II occurred in part because of mechanical picking machines that displaced them. They sought out jobs in iron, steel, and automobile factories, competing with many immigrants for jobs, and engendering their vicious racism. See Anthony Walton, "Technology versus African-Americans," https://www.theatlantic.com/past/docs/issues/99jan/aftech.htm.

8. Veront Satchell, "Innovations in Sugar Cane Mill Technology in Jamaica, 1760–1830," in *Working Slavery, Pricing Freedom: Perspectives from the Caribbean, Africa and the African Diaspora*, ed. Verene Shepherd (Kingston: Ian Randle, 2002), pp. 93–111. Anthony Walton confirmed this view, writing, "[B]lacks have participated as equals in the technological world only as consumers, otherwise existing on the margins of the ethos that defines the nation, underrepresented as designers, innovators, and implementers of our systems and machines. As a group, they have suffered from something that can loosely be called technological illiteracy." He continues, "As the world gets faster and more information-centered, it also gets meaner: disparities of wealth and power strengthen; opportunities change and often fade away." See Anthony Walton, "Technology versus African-Americans."

9. Maria Portuondo, "Plantation Factories: Science and Technology in Late-Eighteenth-Century Cuba," *Technology and Culture* 44, no. 2 (April 2003): 235–36. See also H. A. Gemery and J. S. Hogendorn, "Technological Change, Slavery and the Slave Trade," in *Technology and European Overseas Enterprise: Diffusion, Adaptation, and Adoption*, ed. Michael Adas (New Brunswick, NJ: Rutgers University Press, 1996).

10. Portuondo, "Plantation Factories."

11. Ibid.

12. R. Keith Aufhauser, "Slavery and Technological Change," *Journal of Economic History* 34, no. 1 (1974): 44. He notes that strikes and other disputes with managers also serve as an incentive to lower the wage bill.

13. J. H. Galloway, "Tradition and Innovation in the American Sugar Industry, c. 1500–1800: An Explanation," *Annals of the Association of American Geographers* 75, no. 3 (1985): 341.

14. Aufhauser, "Slavery and Technological Change," p. 42.

15. Lynn Festa, *Sentimental Figures of Empire in Eighteenth-Century Britain and France* (Baltimore: Johns Hopkins University Press, 2006), p. 162.

16. William Cowper, "The Negro's Complaint," http://www.yale.edu/glc/aces/cowper2 .htm.

17. Aufhauser, "Slavery and Technological Change," pp. 46–47.

18. Satchell, *From Plots to Plantations: Land Transactions in Jamaica* (Mona, Kingston, Jamaica: Institute of Social and Economic Research, 1990).

19. See for example David Arnold, *Science, Technology and Medicine in Colonial India* (Cambridge and New York: Cambridge University Press, 2000).

20. Campbell, "Jamaica: The Myth of Economic Development and Racial Tranquility," pp. 16–23.

21. Theo E. S. Scholes, *Sugar and the West Indies* (London: Elliot Stock, 1897), p. 5.

22. Ibid., 18.

23. Ibid.

24. John Soluri, "Bananas Before Plantations. Smallholders, Shippers, and Colonial Policy in Jamaica, 1870–1910," *Iberoamericana* (2001–), Nueva época, año 6, no. 23 (Sep. 2006), pp. 143–59.

25. Michael Salmon, "Land Utilization within Jamaica's Bauxite Land Economy," *Social and Economic Studies* 36, no. 1 (March 1987): 57–92.

26. Sugar Industry Authority of Jamaica, "An Overview of the Sugar Industry of Jamaica," April 2000, http://www.jamaicasugar.org/SIASection/Overview.pdf.

27. Statistical Institute of Jamaica, "Volume of Production of Specified Manufactured Products, 2006–2008," July 18, 2014, http://statinja.gov.jm/productionstats.aspx.

28. "Banana Production," *Science* 17, no. 433 (May 22, 1891): 289–90.

29. A 1932 article in *Economic Geography* as cited in John Soluri, *Banana Cultures: Agriculture, Consumption and Environmental Change in Honduras and the United States* (Austin: University of Texas Press, 2005), p. 62.

30. Jesse Palmer, "The Banana in Caribbean Trade," *Economic Geography* 8, no. 3 (July 1932): 271.

31. Ibid., p. 268.

32. Ibid., p. 272.

33. David Lilienthal, *TVA: Democracy on the March* (New York: Harper and Brothers, 1944).

34. Soluri notes, "The association of bananas with blackness in nineteenth-century Jamaica was sufficiently strong to discourage most white planters from cultivating them for export." See Soluri, "Bananas Before Plantations," pp. 143–59.

35. Ibid.

36. N. W. Simmons, *Bananas* (London: Longmans, 1959), p. 12.

37. Palmer, "The Banana in Caribbean Trade," pp. 262–65, 267–68.

38. Soluri, "Bananas Before Plantations."

39. Geoffrey Evans, "Research and Training in Tropical Agriculture," *Trinidad Journal of the Royal Society of Arts* 87, no. 4499 (Feb. 10, 1939): 336.

40. Michael Adas, *Machines as the Measure of Men: Science, Technology, and Ideologies of Western Dominance* (Ithaca: Cornell University Press, 1989).

41. Evans, "Research and Training in Tropical Agriculture," p. 337.

42. Soluri, "Bananas Before Plantations."

43. "When Is a Banana Ripe?" *British Medical Journal* 1, no. 2265 (May 28, 1904): 1271–72.

44. Bertha M. Wood, "The Banana," *American Journal of Nursing* 28, no. 5 (May 1928): 471–73.

45. Samuel C. Prescott, "The Banana: A Food of Exceptional Value," *Scientific Monthly* 6, no. 1 (Jan. 1918): 65–75.

46. Camilla H. Wedgwood, "Girls' Puberty Rites in Manam Island, New Guinea," *Oceania* 4, no. 2 (Dec. 1933): 132–55.

47. "No Bananas for Babies," *Science News–Letter*, 42, no. 10 (Sep. 5, 1942): 157.

48. Geoffrey Evans, "Research and Training in Tropical Agriculture," *Trinidad Journal of the Royal Society of Arts* 87, no. 4499 (Feb. 10, 1939): 342.

49. "Banana Disease in Fiji," *Bulletin of Miscellaneous Information (Royal Gardens, Kew)* 1892, no. 62 (1892): 48–49.

50. Evans, "Research and Training in Tropical Agriculture," p. 342.

51. Steve Marquardt, "Pesticides, Parakeets, and Unions in the Costa Rican Banana Industry, 1938–1962," *Latin American Research Review* 37, no. 2 (2002): 6.

52. Mitch Lansky, *Beyond the Beauty Strip* (Thomaston, ME: Tilbury House, 1992).

53. Marquardt, "Pesticides, Parakeets, and Unions."

54. For discussion of the CIA involvement in the Guatemalan coup, see David M. Barrett, "Sterilizing a 'Red Infection': Congress, the CIA, and Guatemala, 1954," http://www.cia.gov/library/center-for-the-study-of-intelligence/kent-csi/vol44no5/html/v44i5a03p.htm; and Richard Immerman, *The CIA in Guatemala: The Foreign Policy of Intervention* (Austin: University of Texas Press, 1982).

55. O. Henry, *Cabbages and Kings* (1904), http://www.online-literature.com/o_henry/cabbages-and-kings/1/

56. Ibid.

57. Palmer, "The Banana in Caribbean Trade," p. 267.

58. Clarence F. Jones and Paul C. Morrison, "Evolution of the Banana Industry of Costa Rica," *Economic Geography* 28, no. 1 (Jan. 1952): 6–7.

59. Ibid., pp. 13–16.

60. "Banana Emergency Strikes Costa Rica," Dec. 12, 2013, http://thinkprogress.org/climate/2013/12/12/3056421/costa-rica-banana-emergency/, and "Yellow Peril," Sep. 19, 2013, http://phys.org/news/2013-09-yellow-peril-banana-farms-contaminating.html.

61. Campbell, "Jamaica: The Myth of Economic Development and Racial Tranquility," pp. 16–23.

62. Michael Salmon, "Land Utilization within Jamaica's Bauxite Land Economy," *Social and Economic Studies* 36, no. 1 (March 1987): 57–92.

63. Wilfred Brining, "Bauxite and Aluminium with Particular Reference to the Commonwealth," *Journal of the Royal Society of Arts* 109, no. 5063 (Oct. 1961): 877–89.

64. Cynthia L. Ogden, Brian K. Kit, Margaret D. Carroll, and Sohyun Park, "Consumption of Sugar Drinks in the United States, 2005–2008," NCHS Data Brief, no. 71, Aug. 2011, http://www.cdc.gov/nchs/data/databriefs/db71.htm.

65. http://www.banthebottle.net/bottled-water-facts/.

66. Salmon, "Land Utilization within Jamaica's Bauxite Land Economy."

67. For discussion of the way governments have viewed people and nature in the modernizing process, and the impact of a kind of sociological aggregation of the masses as ways to treat them, see James Scott, *Seeing Like a State* (New Haven, CT: Yale University Press, 1999).

68. John H. Bounds, "Restoration of Mined Land to Farming in Jamaica," *Revista Geográfica* 80 (June 1974): 105–10.

69. Brining, "Bauxite and Aluminium," p. 887.

70. Salmon, "Land Utilization within Jamaica's Bauxite Land Economy."

71. Ibid.

72. W. A. Snaith, "Forestry Development by a Bauxite Mining Company," *Commonwealth Forestry Review* 52, no. 1 (151) (March 1973): 79–81.

73. Salmon, "Land Utilization within Jamaica's Bauxite Land Economy."

74. Bounds, "Restoration of Mined Land."

75. Ibid.

76. Jamaica Bauxite Institute, "Development of the Bauxite/Alumina Sector," http://www.jbi.org.jm/pages/industry, accessed Sep. 20, 2014.

77. "Local Sugar Production Up—$154.3m Loan Provides Lifeline for Industry," *Jamaica Gleaner*, Nov. 8, 2011, http://jamaica-gleaner.com/gleaner/20111108/lead/lead2.html. See also Roger Clarke, "Minister's Speech," Oct. 4, 2012, http://www.moa.gov.jm

/Speeches/2012/20121004_Minister's_speech_%20at_the_Sugar%20transformation%20 review%20seminar.php.

78. http://www.jamaicasugar.org/FactoryHistory/FactoryHistory.html, accessed Sep. 20, 2014, and Crop Site, "Jamaica Sugar Annual Report 2013," June 24, 2013, http:// www.thecropsite.com/reports/?id=2274.

79. Dan Koeppel, "Yes, We Have No Bananas," June 18, 2008, http://www.nytimes .com/2008/06/18/opinion/18koeppel.html?_r=0. See his *Banana: The Fate of the Fruit That Changed the World* (New York: Penguin, 2007).

80. Steve Connor, "Lack of Sex Life Threatens Banana Crops," *Independent*, July 27, 2001, http://news.nationalgeographic.com/news/2001/07/0726_wirebanana.html.

81. Pablo Neruda, "United Fruit Company," http://faculty.sfhs.com/lesleymuller/world _history/latin_America/ufc.pdf.

CHAPTER 4: Mass-Produced Nutrition

1. FDA, "Food Irradiation: What You Need to Know," March 7, 2014, http://www .fda.gov/Food/ResourcesForYou/Consumers/ucm261680.htm.

2. Richard White, *The Organic Machine* (New York: Hill and Wang, 1995). On Hanford and its human and environmental costs, see Kate Brown, *Plutopia* (New York: Oxford University Press, 2013). My thanks to Jennifer Alexander for introducing me to White's book and more importantly to the history of potatoes in Washington.

3. Phillip B. C. Jones, "Revolution on a Dare: Edmund Smith and His Famous Fish-Butchering Machine," *Columbia*, 20, no. 3 (2006): 37–42.

4. "Hydropower to Make the American Dream Come True (1939)," http://www.bpa .gov/news/AboutUs/History/Pages/Hydro-Power-to-Make-the-American-Dream -Come-True.aspx.

5. "Columbia Basin Project," http://www.usbr.gov/projects/Project.jsp?proj_Name =Columbia+Basin+Project.

6. USDA, "Potatoes," http://www.ers.usda.gov/topics/crops/vegetables-pulses/potatoes .aspx#.U6vYe_ldVaU. See also US Potato Board, *National Potato Council 2013 Potato Statistical Yearbook* (2013), http://nationalpotatocouncil.org/files/4613/6940/5509/2013 _NPCyearbook_Web_FINAL.pdf.

7. "History of Potatoes," http://www.potatoes.com/our-industry/history/.

8. B. Delworth Gardner and Ray G. Huffaker, "Cutting the Loss from Federal Irrigation Subsidies," *Choices* (third quarter 1993), p. 15, http://ageconsearch.umn.edu /bitstream/130477/2/DelworthandHuffaker.pdf.

9. McCain USA, "McCain: About Us," http://www.mccainusa.com/aboutUs.aspx.

10. "Dehydrated Potatoes," http://www.potatoesusa.com/products.php?sec=Dehydrated %20Potatoes.

11. Garth Taylor, Paul Patterson, Joe Guenthner, and Lindy Widner, *Contributions of the Potato Industry to Idaho's Economy* (University of Idaho: CIS, Oct. 2007), http://www .cals.uidaho.edu/edcomm/pdf/cis/cis1143.pdf. The workers' income taxes may help Idaho restore cuts to thousands of disabled people who were removed from welfare in the past few years.

12. See "Export Market," http://www.uspotatoes.com/downloads/Frozen%20Potato% 20Products-Export_NEW.pdf.

13. Seattle Labor Civil Rights and History Project, "Timeline: Farm Worker Orga- nizing in Washington State," http://depts.washington.edu/civilr/farmwk_timeline .htm.

14. Thomas Arcury et al., "Repeated Pesticide Exposure Among North Carolina Migrant and Seasonal Workers," *American Journal of Industrial Medicine*, 53, no. 8 (Aug. 2010): 802–13, http://www.ncbi.nlm.nih.gov/pmc/articles/PMC2904622/.

15. Washington State Potato Commission, *Teaching Taters*, http://www.potatoes.com. See also such fun children's books as Jennifer Julius's *I Like Potatoes*, *Potatoes on Tuesday* (1995) and Toby Speed and Barry Root's *Brave Potatoes* (2000).

16. Willian Cronon, *Nature's Metropolis* (New York: W. W. Norton, 1991).

17. "Corn. Background," http://www.ers.usda.gov/topics/crops/corn/background .aspx#.U1PnOlVdWKI, and "Beef Industry Statistics," http://www.beefusa.org/beef industrystatistics.aspx.

18. David Biello, "Intoxicated on Independence: Is Domestically Produced Ethanol Worth the Cost?" *Scientific American*, July 28, 2011, http://www.scientificamerican.com /article/ethanol-domestic-fuel-supply-or-environmental-boondoggle/.

19. EWG Farm Subsidies, "Corn Subsidies in the United States Totaled $84.4 billion from 1995–2012," http://farm.ewg.org/progdetail.php?fips=00000&progcode =corn.

20. E. C. Pasour, "The Farm Bill Lavishes Cash on Upper-Income Farmers," Nov. 1, 2008, http://www.fee.org/the_freeman/detail/us-agricultural-programs-who-pays.

21. Gail Hollander, *Raising Cane in the 'Glades* (Chicago: University of Chicago Press, 2008).

22. See for example General Accounting Office, "Foreign Farm Workers in US: Department of Labor Action Needed to Protect Florida Sugar Cane Workers," HRD-92-95 (June 30, 1992), http://www.gao.gov/products/HRD-92-95.

23. Robert Coker (senior vice president, US Sugar), "Problems with River and Lagoon Long Predate Sugar Farming," Jan. 16, 2014, http://www.ussugar.com/press_room/releases /ProblemsRiverandLagoon.pdf.

24. Clay Landry, "Who Drained the Everglades?," http://perc.org/articles/who-drained -everglades.

25. Michael Grunwald, "Booting US Sugar From the Everglades," June 24, 2008, http://content.time.com/time/nation/article/0,8599,1817390,00.html.

26. David Guest, "Sugar Industry Seeks to Use Everglades as Toilet," http:// earthjustice.org/blog/2012-september/sugar-industry-seeks-to-use-everglades-as-toilet.

27. "Florida Everglades," http://www.nrdc.org/water/conservation/qever.asp#industry.

28. Corn Products International Product Applications, http://www.cornproducts .com?productApplications.shtml.

29. Mark Lesney, "Agricultural and Food Chemicals," http://pubs.acs.org/supplements /chemchronicles2/pdf/023.pdf.

30. See Edmund Russell, *War and Nature* (Cambridge: Cambridge University Press, 2001), for the rise of the battle against pests and the lexicon of war—and chemicals—used in that battle.

31. See *Agribusiness Examiner*, Jan. 30, 2003, Issue 220, http://www.electricarrow.com /CARP/agbiz/220.htm.

32. John Greenwald, "The Fix Was in at ADM," *Time Magazine* 148, no. 20 (Oct. 28, 1996), http://www3.nd.edu/~mgrecon/datafiles/articles/admpricefixing.html.

33. Robert Manor, "ADM Settles Price-Fixing Charges for $400 Million," *Chicago Tribune*, June 19, 2004, http://articles.chicagotribune.com/2004-06-19/business/0406190182 _1_lysine-and-citric-acid-mark-whitacre-corn-syrup.

34. Cargill, "Our History," http://www.cargill.com/company/history/index.jsp, accessed September 21, 2014.

35. Corn Refiners Association, "A Brief History of the Corn Refining Industry," http://www.corn.org/about-2/history/#sthash.DTI1583B.dpuf.

36. "Tapping the Treasure," http://www.corn.org/wp-content/uploads/2009/12/Tapping.pdf.

37. Marion Nestle, http://www.foodpolitics.com/tag/hfcs-high-Fructose-corn-syrup/.

38. Iowa Corn Promotion Board, "Iowa Corn," https://www.iowacorn.org/, and Illinois Corn Growers' Association, "Illinois Corn," http://www.ilcorn.org/home.

39. Iowa Corn Promotion Board, "History," http://www.iowacorn.org/news/news_8.html.

40. Kentucky Corn Growers Association, "Corn Is All Around Us!" for fourth and fifth graders, http://www.kycorn.org/corneducation/presentation.html, accessed on 4/5/2005.

41. Anna Lappe, "Don't Sugar-Coat High-Fructose Corn Syrup," Sep. 20, 2010, http://www.theatlantic.com/health/archive/2010/09/dont-sugar-coat-high-Fructose-corn-syrup/63195/.

42. W. H. Glinsmann, H. Irausquin, and Y. K. Park, "Evaluation of Health Aspects of Sugars Contained in Carbohydrate Sweeteners: Report of Sugars Task Force, 1986," *Journal of Nutrition* 116 (1986): S1–S216; Glinsmann and B. A. Bowman, "The Public Health Significance of Dietary Fructose," *American Journal of Clinical Nutrition* 58 (1993): 820–23; Office of the Federal Register, National Archives and Records Administration, *Federal Register* 48 (1983): 5715–19; S. Reiser et al., "Indices of Copper Status in Humans Consuming a Typical American Diet Containing Either Fructose or Starch," *American Journal of Clinical Nutrition* 42 (1985): 242–51; and J. Hallfrisch et al., "Plasma Fructose, Uric Acid, and Inorganic Phosphorus Responses of Hyperinsulinemic Men Fed Fructose," *Journal of American College of Nutrition* 5 (1986): 61–68.

43. J. Hallfrisch et al., "Blood Lipid Distribution of Hyperinsulinemic Men Consuming Three Levels of Fructose," *American Journal of Clinical Nutrition* 37 (1983): 740–48, and G. A. Bray et al., "Consumption of High-Fructose Corn Syrup in Beverages May Play a Role in the Epidemic of Obesity," *American Journal of Clinical Nutrition* 79 (2004): 537–43.

44. John L. Sievenpiper et al., "Effect of Fructose on Body Weight in Controlled Feeding Trials: A Systematic Review and Meta-analysis," *Annals of Internal Medicine* 156, no. 4 (2012): 291–304.

45. Bray et al., "Consumption of High-Fructose."

46. John S. White, "Straight Talk About High-Fructose Corn Syrup: What It Is and What It Ain't," *American Journal of Clinical Nutrition* 88, no. 6 (Dec. 2008): 1716s–21s, http://ajcn.nutrition.org/content/88/6/1716S.full.

47. "Sugar Industry CEO Says There's No Link Between Sugar and Obesity," Aug. 22, 2005, http://www.naturalnews.com/011190.html#ixzz36Dzg3c00.

48. Like anti-communist public health conspiracy adherents in the 1950s and 1960s, Citizens for Health also opposes fluoridated water as a big government conspiracy. See "Citizens for Health Education Foundation," http://www.citizens.org/foundation/.

49. For a series of links to syrupy responses to corny lawsuits, see "Search Results for Food and Drug Administration," http://sugar.org/cra-lawsuit/?s=Food+and+Drug+Administration, accessed Sep. 21, 2014.

50. Derek Hunter, "You Can't Sugarcoat the Sugar Industry's Greed," Jan. 31, 2014, http://dailycaller.com/2014/01/31/you-cant-sugarcoat-the-sugar-industrys-greed/#ixzz36DA1dOw5.

51. Susan Salisbury, "Sugar-Corn Syrup Court Fight Reveals Inner Workings," *Palm Beach Post*, Feb. 14, 2014, http://www.gosanangelo.com/news/2014/feb/14/sugar-corn

-syrup-court-fight-reveals-inner/?print=1. The plaintiffs are: Western Sugar Cooperative, Michigan Sugar Co., C & H Sugar Co. Inc., U.S. Sugar Corp., American Sugar Refining Inc., Amalgamated Sugar Co. LLC, Imperial Sugar Co., Minn-Dak Farmers Cooperative, American Sugar Cane League U.S.A. Inc., and Sugar Association Inc. The defendants are: Corn Refiners Association Inc., Archer-Daniels-Midland Co., Cargill Inc., Corn Products International Inc., Roquette America Inc., and Tate & Lyle Ingredients Americas Inc.

52. Center for Consumer Freedom, "What Is the Center for Consumer Freedom," http://www.consumerfreedom.com/about/, accessed Sep. 21, 2014.

53. David Knowles, "Sugar Industry Denies Misleading Public Despite Media Reports Exposing 'Pay-for-Play' Campaign Against Corn Refiners," Oct. 30, 2012, http://www.corn.org/press/newsroom/sugar-industry-pay-for-play-reports/#sthash.QnsD1K5C.dpuf.

54. Eric Lipton, "Rival Industries Sweet-Talk the Public," Feb. 12, 2014, http://www.nytimes.com/2014/02/12/business/rival-industries-sweet-talk-the-public.html?ref=business&_r=0.

55. Marion Nestle, "Guess Who Funded the Contradictory Fructose Study?" http://www.foodpolitics.com/tag/hfcs-high-Fructose-corn-syrup/.

56. For an early effort to understand technical disagreements, see Allan Mazur, *The Dynamics of Technical Controversy* (Washington, DC: Communications Press, 1981). For a contemporary study that considers empirical and social foundations of disputes, see Dominique Pestre, *A Contre-science: Politiques et Savoirs des Sociétés Contemporaines* (Paris: Editions du Seuil, 2013).

57. Naomi Oreskes and Erik Conway, *Merchants of Doubt* (New York: Bloomsbury Press, 2010).

58. Allan Brandt, *The Cigarette Century* (New York: Basic Books, 2007).

59. Daniel Puzo, "The Battle of the Bureaucrats: Agencies: Rivals USDA and FDA," *Los Angeles Times*, May 30, 1991, http://articles.latimes.com/1991-05-30/food/fo-3454_1_food-safety. See also General Accounting Office, *Food Safety: More Disclosure and Data Needed to Clarify Impact of Changes to Poultry and Hog Inspections*, GAO-13-775 (Washington, DC: GAO, 2013).

60. Shaya Tayefe Mohajer, "FDA Creating New Nutrition Facts Label," Sep. 3, 2011, http://www.huffingtonpost.com/2011/09/03/fda-nutrition-facts_n_948075.html.

61. See David Kessler, *A Question of Intent: A Great American Battle with a Deadly Industry* (New York: Public Affairs, 2001).

62. Bob Herbert, "That's Where the Money Is," Oct. 5, 2010, http://www.nytimes.com/2010/10/05/opinion/05herbert.html?_r=0.

63. David Kessler, *The End of Overeating: Taking Control of the Insatiable American Appetite* (New York: Public Affairs, 2001). See also http://articles.washingtonpost.com/2009-04-27/news/36916774_1_ingredient-labels-kessler-food-industry.

64. Matthias Schulze et al., "Sugar-Sweetened Beverages, Weight Gain and Incidence of Type 2 Diabetes in Young and Middle-Aged Women," *Journal of the American Medical Association* 292, no. 8 (Aug. 2004): 927–34.

65. Michael I. Gora, Stanley J. Ulijaszek, and Emily E. Ventura, "High Fructose Corn Syrup and Diabetes Prevalence: A Global Perspective," *Global Public Health* 8, no. 1 (2013): 55–64.

66. Hella Jurgens et al., "Consuming Fructose-sweetened Beverages Increases Body Adiposity in Mice," *Obesity* 13, no. 7 (July 2005): 1146–56, and Ming Song et al., "Modest Fructose Beverage Intake Causes Liver Injury and Fat Accumulation in Marginal Copper Deficient Rats," *Obesity* 21, no. 8 (Aug. 2013): 1669–75.

67. Ron Nixon, "Congress Blocks New Rules on School Lunches," Nov. 15, 2011, http://www.nytimes.com/2011/11/16/us/politics/congress-blocks-new-rules-on-school -lunches.html?_r=0.

68. Diane Pratt-Heavner, "Myth vs. Fact on Healthy, Hunger-Free Kids Act School Meals Implementation," May 22, 2014, http://www.schoolnutrition.org/Blog2.aspx?id =20479&blogid=564&terms=Healthy+Hunger-Free+Kids+Act.

69. See *Nutrition and Your Health: Dietary Guidelines for Americans: 2005 DGAC Report,* http://www.health.gov/dietaryguidelines/dga2005/report/HTML?G1_Glossary .htm, accessed April 5, 2005, that includes glossary of terms such as "added sugars" that do not include naturally occurring sugars, "complex carbohydrates," "foodborne diseases," "HFCS," and "whole-grain foods."

70. "Corn Syrup Rebranding as 'Corn Sugar,' " http://www.deliciousmusings.com/?p =6975, and http://www.cdc.gov/nutrition/everyone/fruitsvegetables/index.html?s_cid =tw_ob191.

71. Anne Lappe, "Don't Sugar Coat High-Fructose Corn Syrup," Sep. 20, 2010, http://www.theatlantic.com/health/archive/2010/09/dont-sugar-coat-high-Fructose -corn-syrup/63195/.

72. See USDA, "Food and Nutrition," Aug. 27, 2014, http://www.usda.gov/wps/portal /usda/usdahome?navid=food-nutrition.

73. US Department of Agriculture, "Agriculture Secretary Vilsack Highlights New 'Smart Snacks in School' Standards; Will Ensure School Vending Machines, Snack Bars Include Healthy Choices," June 13, 2013, http://www.usda.gov/wps/portal/usda/usda home?contentid=2013/06/0134.xml.

CHAPTER 5: Technology and (Natural) Disasters

1. M. A. Younus, R. D. Bedford, and M. Morad, "Not So High and Dry: Patterns of 'Autonomous Adjustment' to Major Flooding Events in Bangladesh," *Geography* 90, no. 2 (Summer 2005): 112–20.

2. For the "racism" of Robert Moses, the father of this parkway system, see Robert Caro, *The Power Broker* (New York: Knopf, 1974).

3. "More Flooding on the Bronx River Parkway," March 7, 2013, http://scarsdale10583 .com/201103071460/today-s-news/bronx-river-floods-parkway.html.

4. "Bronx River Parkway Historical Overview," http://www.nycroads.com/roads /bronx-river/.

5. John McPhee, *The Control of Nature* (New York: Farrar, Straus, Giroux, 1989).

6. Pete Daniel, *Deep'n As It Comes: The 1927 Mississippi River Flood* (New York: Oxford University Press, 1977), and John Barry, *Rising Tide: The Great Mississippi Flood and How It Changed America* (New York: Simon & Schuster, 1997).

7. Craig Colton, *Perilous Place and Powerful Storms: Hurricane Protection in Coastal Louisiana* (Jackson: University Press of Mississippi, 2009), and *An Unnatural Metropolis: Wresting New Orleans from Nature* (Baton Rouge: Louisiana State University Press, 2005).

8. For a discussion of race and natural disasters, see Jason David Rivera and DeMond Shondell Miller, "Continually Neglected: Situating Natural Disasters in the African American Experience," *Journal of Black Studies* 37, no. 4, Katrina: Race, Class, and Poverty (Mar. 2007): 502–22.

9. Morgan Whitaker, "Former FEMA Director Michael Brown Criticizes Obama for Responding too [sic] Sandy 'Quickly,' " Oct. 30, 2012, http://www.msnbc.com/politicsnation /former-fema-director-michael-brown-criticizes.

10. Elizabeth Zimmerman, FEMA assistant administrator, "Disaster Assistance Fact Sheet DAP9580.8," Oct. 1, 2009, http://www.fema.gov/pdf/government/grant/pa/9580_8.pdf.

11. "State of the Beach/State Reports/DE/Beach Fill," http://www.beachapedia.org/State_of_the_Beach/State_Reports/DE/Beach_Fill. According to a 1962 Federal law, "Section 103 of the 1962 River and Harbors Act provides authority for the Corps of Engineers to develop and construct projects to protect the shores of publicly owned property by constructing revetments, groins, jetties, to include periodic sand replacement. Each project is limited to a federal cost of not more than $3 million." See http://www.mvn.usace.army.mil/pd/projectslist/home.asp?projectID=73.

12. "Chris Christie Touts Funding Support for the Jersey Shore During Tour of State Beaches," Aug. 24, 2011, http://jerseyshore.surfrider.org/2011/08/29/chris-christie-touts-funding-support-for-the-jersey-shore-during-tour-of-state-beaches/.

13. Robert Young, "A Beach Project Built on Sand," Aug. 21, 2014, http://www.nytimes.com/2014/08/22/opinion/a-beach-project-built-on-sand.html?action=click&contentCollection=Opinion&region=Footer&module=MoreInSection&pgtype=article.

14. "Sea Level Rise," http://ocean.nationalgeographic.com/ocean/critical-issues-sea-level-rise/.

15. Clare Murphy, "Starting from Scratch in Bam," Jan. 2, 2004, http://news.bbc.co.uk/2/hi/middle_east/3363125.stm.

16. Edward Wong, "A Chinese School, Shored Up By Its Principal, Survived Where Others Fell," *New York Times*, Sunday, June 15, 2008; and Geoffrey York, "Why China's Buildings Crumbled: Survivors Blame Corruption, Shoddy Construction and Cost Cutting," May 15, 2008, http://v1.theglobeandmail.com/servlet/story/RTGAM.20080515.wchinaside15/front/Front/Front/.

17. Richard Kerr, "How the Armenian Quake Became a Killer," *Science* 243, no. 4888 (Jan. 13, 1989), p. 170.

18. Steven Buckhingham, "The Flaming Rat Case: A Revisionist Analysis," June 27, 2012, http://abnormaluse.com/2012/06/the-flaming-rat-case-a-revisonist-analysis.html.

19. See Ravi Rajan, "Disaster, Development and Governance: Reflections on the 'Lessons' of Bhopal," *Environmental Values* 11, no. 3, Science, Development and Democracy (Aug. 2002): 369–94.

20. Oil Spill Case Histories 1967–1991, Hazardous Materials Response and Assessment Division, Report No. HMRAD 92-11 (Seattle: National Oceanic and Atmospheric Administration, Sep. 1992).

21. EPA, "Oil Pollution Act Overview," September 8, 2014, http://www.epa.gov/oem/content/lawsregs/opaover.htm.

22. International Maritime Organization, "Construction Requirements for Oil Tankers," http://www.imo.org/OurWork/Environment/PollutionPrevention/OilPollution/Pages/constructionrequirements.aspx.

23. William R. Freudenberg and Robert Gramling, *Blowout in the Gulf: The BP Oil Spill Disaster and the Future of Energy in America* (Cambridge, MA: MIT Press, 2011).

24. Ibid.

25. On the Japanese government–commissioned study that documents the failures of the nuclear industry to secure safety, see "Fukushima Accident—Disaster Response Failed—Report," Dec. 26, 2011, http://www.bbc.co.uk/news/world-asia-16334434. See http://www.hse.gov.uk/nuclear/fukushima/interim-report.pdf. for the British government response to the lessons of Fukushima.

26. "Walking on a Tightrope: Water Contamination by Nuclear Fallout," *Asahi*, March 11, 2014. Translated by Colby student Melissa Meyer, '16.

27. "Increasing Fear About Children's Thyroids—Cancer Test: Fumbling without Precedent," *Asahi*, March 8 2014. Translated by Colby student Melissa Meyer, '16.

28. Langdon Winner, *The Whale and the Reactor* (Chicago: University of Chicago Press, 1986), and Richard Meehan, *The Atom and the Fault* (Cambridge, MA: MIT Press, 1984).

29. http://www.sanluisobispo.com/2011/03/14/1521331/diablo-canyon-nuclear-tsunami.html.

30. Kseniya Yershova and Mary Wells, "Armenia Reopening Metsamor," http://www.nti.org/db/nisprofs/armenia/bkgndrep.htm.

31. John Daly, "Armenia's Aging Metsamor Nuclear Power Plant Alarms Caucasian Neighbors," Oct. 3, 2011, http://oilprice.com/Alternative-Energy/Nuclear-Power/Armenias-Aging-Metsamor-Nuclear-Power-Plant-Alarms-Caucasian-Neighbors.html.

32. "Metsamor Nuclear Power Plant Would Withstand Japan Earthquake: Expert," March 21, 2011, http://www.epress.am/en/2011/03/21/metsamor-nuclear-power-plant-would-withstand-japan-earthquake-expert.html.

33. See J. Samuel Walker, *Three Mile Island: A Nuclear Crisis in Historical Perspective* (Berkeley: University of California Press, 2004).

34. Georgy Bovt, "Putin's Vertical Power Integration," Aug. 13, 2010, http://www.themoscowtimes.com/opinion/article/putins-vertical-power-disaster/412296.html.

35. Paul Goble, "Analysts: Putin's Destruction of Forest Service in 2007 Behind Russia's Current Fire Disaster," Aug. 9, 2010, http://www.kyivpost.com/news/opinion/op_ed/detail/77509/#ixzz1aZZxgKWC.

36. Gleb Bryanski, "Opposition Says Putin Law Cripples Fire-Fighting," Aug. 4, 2010, http://uk.reuters.com/article/2010/08/04/uk-russia-fires-law-idUKTRE6723BL20100804.

37. "SC Forestry Commission Archives Fire Prevention Posters," http://www.state.sc.us/forest/posters.htm.

38. For a history of the interaction between humans and fire, including fire-fighting and suppression, see Stephen J. Pyne, *Fire: Nature and Culture* (London: Reaktion Books, 2012).

39. "Forest Fires," http://www.enviroliteracy.org/article.php/46.html.

40. H. T. Gisborne, "Forest Pyrology," *Scientific Monthly* 49, no. 1 (Jul. 1939): 21–30, and Herbert C. Hanson, "Fire in Land Use and Management," *American Midland Naturalist* 21, no. 2 (Mar. 1939): 415–34.

41. National Park Service, "Visiting the Hoh Rain Forest," http://www.nps.gov/olym/planyourvisit/visiting-the-hoh.htm.

42. US Forest Service, "US Forest Service Fire Prevention and Control," http://www.foresthistory.org/ASPNET/Policy/Fire/Prevention/Prevention.aspx.

43. Warren Dean, *With Broadax and Firebrand* (Berkeley: University of California Press, 1997), and A. D. Nobre, *The Future of Climate in Amazonia* (Cuiaba: Articulacion Regional Amazonica, 2014).

44. John McPhee, "Los Angeles Against the Mountains," in *The Control of Nature*.

45. Friends of the River Narmada, "The Sardar Sarovar Dam: A Brief Introduction," May 10, 2006, http://www.narmada.org/sardarsarovar.html.

46. Arundhati Roy, "The Greater Common Good," http://www.narmada.org/gcg/gcg.html.

47. "Electrobras," http://www.eletrobras.com/elb/data/Pages/LUMISB33DBED6ENIE.html.

48. P. M. Fearnside, "Environmental Impacts of Brazil's Tucuruí Dam: Unlearned Lessons for Hydroelectric Development in Amazonia," *Environmental Management* 27, no. 3 (2001): 377–96.

49. Alfred Crosby, *Ecological Imperialism* (Cambridge; New York: Cambridge University Press, 2004).

50. I discuss "Corridors of Modernization" in *Industrialized Nature* (Washington, DC: Island Press, 2002), ch. 4.

51. T. Leino and M. Lodenius, "Human Hair Mercury Levels in Tucuruí Area, State of Pará, Brazil," *Science of the Total Environment* 175, no. 2 (2005): 119–25.

52. See Bill Luckin, "Nuclear Meltdown and the Culture of Risk," *Technology and Culture* 46, no. 2 (April 2005): 393–99; and Greg Bankoff, "No Such Thing as Natural Disasters," Aug. 23, 2010, http://hir.harvard.edu/no-such-thing-as-natural-disasters.

53. Jim Yardley, "Report on Deadly Factory Collapse in Bangladesh Finds Widespread Blame," May 22, 2013, http://www.nytimes.com/2013/05/23/world/asia/report-on -bangladesh-building-collapse-finds-widespread-blame.html.

54. Ulrich Beck, "The Anthropological Shock: Chernobyl and the Contours of the Risk Society," *Berkeley Journal of Sociology* 32 (1992): 153–65.

55. Ibid., and *Risk Society: Towards a New Modernity* (London: Sage, 1992). On environmental racism and the inequity of risk, see United Church of Christ, *Toxic Wastes and Race in the United States* (New York: UCC 1987), and Robert Bullard, "Environmental Justice in the 21st Century: Race Still Matters," *Phylon* 49, no. 3/4 (Autumn–Winter 2001): 151–71.

56. Bankoff, "No Such Thing as Natural Disasters."

57. Ibid.

CHAPTER 6: Big Artifacts

1. Arnold Pacey, *The Maze of Ingenuity: Ideas and Idealism in the Development of Technology* (Cambridge, MA: MIT Press, 1976), and David Landes, *Revolution in Time: Clocks and the Making of the Modern World* (Cambridge, MA: Harvard University Press, 2000).

2. Royal Commission, *Official Catalogue of the Great Exhibition of the Works of Industry of All Nations, 1851* (London: Spicer Brothers, 1851), and Centennial Board of Finance, *Visitors' Guide to the Centennial Exhibition and Philadelphia. May 10th to November 10th, 1876* (Philadelphia: J.B. Lippincott & Co., 1876).

3. Albert Speer, *Inside the Third Reich* (New York: Avon, 1970), and Martino Stierli, "Building No Place: Oscar Niemeyer and the Utopias of Brasilia," *Journal of Architectural Education* 67, no. 1 (2013): 8–16.

4. John F. Kennedy, "Speech to Joint Session of the US Congress," May 25, 1961, http://www.jfklibrary.org/Asset-Viewer/Archives/JFKWHA-032.aspx.

5. See, for example, Thomas Hughes, *Networks of Power: Electrification in Western Society, 1880–1930* (Baltimore: Johns Hopkins University Press, 1983), and *American Genesis: A Century of Invention and Technological Enthusiasm, 1870–1970* (New York: Viking, 1989).

6. For discussion of the European contribution to the space race, see John Krige and Arturo Russo, *Europe in Space, 1960–1973* (Noordwijk: European Space Agency Publications Division, 1994), and *A History of the European Space Agency. The History of ESRO and ELDO from 1958 to 1973*, vol. 1 (Noordwijk: ESA SP1235, 2000); also see Krige, Russo, and L. Sebesta, *A History of the European Space Agency. The History of ESA from 1973 to 1987*, vol. 2 (Noordwijk: ESA SP1235, 2000). On nuclear ideologies, see for example Gabrielle Hecht, *The Radiance of France* (Cambridge, MA: MIT Press, 1998).

7. Among the many studies that consider the ideological and social aspects of large-scale technologies, see Walter A. McDougall, . . . *the Heavens and the Earth: A Political History of the Space Age* (New York: Basic Books, 1985); Stephen Kotkin, *Magnetic Mountain* (Berkeley: University of California Press, 1995); Thomas Zeller, *Driving Germany* (New York: Berghahn Books, 2007); Alf Nilsen, *Dispossession and Resistance in India* (London, New York: Routledge, 2010); and Mark Reisner, *Cadillac Desert* (New York: Viking, 1986).

8. Paul Josephson, "'Projects of the Century' in Soviet History: Large Scale Technologies from Lenin to Gorbachev," *Technology and Culture* 36, no. 3 (July 1995): 519–59.

9. Loren Graham has written extensively about the political and cultural context of technology in Russian society, for example in *The Ghost of the Executed Engineer* (Cambridge, MA: Harvard University Press, 1993), and his recent *Lonely Thoughts* (Cambridge, MA: MIT Press, 2014).

10. Ivan Papanin, *Zhizn' na L'dine. Dnevnik* (Moscow: Pravda, 1938); M. Vodop'ianov, *Poliarnyi Letchik. Rasskazy* (Leningrad: Leningradskoe Gazetno-zhurnal'noe i Knizhnoe Izdatel'stvo, 1954).

11. See Kendall Bailes, "Technology and Legitimacy: Soviet Aviation and Stalinism in the 1930s," *Technology and Culture* 17 (1976): 55–81, and Scott Palmer, *Dictatorship of the Air* (Cambridge: Cambridge University Press, 2006).

12. V. K. Buinitskii, *812 Dnei v Dreifuiushchikh lLdakh* (Moscow: Glavsevmorput, 1945), pp. 15–23.

13. "Early Icebreakers," http://www.globalsecurity.org/military/world/russia/icebreaker -1.htm, accessed Sep. 22, 2014.

14. Putin addressed precisely these issues in his candidate thesis that he defended at the St. Petersburg Mining Institute in 1997. He argued that Russia's great natural resources were the key to remaking the country into a great economic power with a high standard of living based on the "fatherland's processing industry based on the extractive complex." See Harley Balzer, "Vladimir Putin's Academic Writings and Russian Natural Resource Policy," *Problems of Post-Communism* (Jan./Feb. 2006): 48–54.

15. On Soviet Arctic exploration see John McCannon, *Red Arctic* (New York: Oxford University Press, 1998) for the 1930s, and Paul Josephson, *The Conquest of the Russian Arctic* (Cambridge, MA: Harvard University Press, 2014). The Arctic has become transformed from a site of Cold War competition to one of fierce economic competition, and now perhaps again to military competition. See Lassi Heininen (with Heather Nicol), "The Importance of Northern Dimension Foreign Policies in the Geopolitics of the Circumpolar North," *Geopolitics* 12, no. 1 (Feb. 2007): 133–65.

16. President of Russia, "Maritime Doctrine of Russian Federation 2020," July 27, 2001, http://www.oceanlaw.org/downloads/arctic/Russian_Maritime_Policy_2020.pdf.

17. President of Russia, "The Basics of State Policy of the Russian Federation in the Arctic Region," Sep. 18, 2008, http://img9.custompublish.com/getfile.php/1042958.1529 .avuqcurreq/Russian+Strategy.pdf?return=www.arcticgovernance.org, and http://www .arcticgovernance.org/russia-basics-of-the-state-policy-of-the-russian-federation-in-the -arctic-for-the-period-till-2020-and-for-a-further-perspective.4651232-142902.html. Russian leaders insisted years ago that they would accomplish these goals in the spirit of international cooperation through the maintenance of mutually advantageous bilateral and multilateral agreements and treaties, and through the sharing of information about the Arctic zone. But since the annexation of Crimea, Russian leaders have become more bellicose about the Arctic zone as well. For discussion of the relation between

politics, investment, gas, and oil, see Per Hogselius, *Red Gas* (New York: Palgrave Macmillan, 2013). See also Marshall Goldman, *Petrostate* (New York: Oxford University Press, 2009).

18. President of Russia, "Security Council Meeting on Shipbuilding Development," June 9, 2010, http://eng.kremlin.ru/news/399.

19. President of Russia, "Executive Order on Implementing Plans for Developing Armed Forces and Modernising Military-Industrial Complex," May 7, 2012, http://eng .kremlin.ru/news/3777.

20. Evgenii Beliakov, "Arkticheskie Vezdekhody," Oct. 8, 2011, http://kp.ru/daily/25767 /2751896/.

21. Andrei Ozharovskii, "Lozh' na Pervom: Po Mneniiu Zhurnalistov na Ledkole na Ledokole Lenin Avarii Ne Bylo," May 6, 2009, http://www.bellona.ru/weblog/andrey -ozharovsky/1241708613.5. For a map of other reactors dumped in the Arctic Ocean, see http://www.solovki.ca/danger/radiation_02.php.

22. Nataliia Antopkina, "Rossiia Prazdnuet 35-letie Pokoreniia Severnogo Poliusa Ledokolom 'Arktika,'" Aug. 12, 2012, http://www.kp.ru/online/news/1224536/, Aug. 16, 2012. Aerial research at the Arctic and Antarctic Research Institute on the Arctic ice regime facilitated the North Pole journey of the *Arktika*.

23. "Ledokhol 'Arktika' Otpravili na Pensiia," Oct. 3, 2008, http://kp.ru/online/news /148296/.

24. Trude Pettersen, "New Icebreakers Could Be Built Abroad," Dec. 12, 2013, http://barentsobserver.com/en/arctic/2013/12/new-icebreakers-could-be-built-abroad -12-12, and Yuri Golotuik, "Safeguarding the Arctic," in *Russia in Global Affairs*, Aug. 9, 2008, http://eng.globalaffairs.ru/person/p_1887.

25. "Est' Podozreniia, Chto Proekt Bol'she Sviazan s Osvoeniem Deneg, Chem s Polucheniem Rezultata," Aug. 24, 2012, http://kommersant.ru/doc/2007338?isSearch =True. The new *Arktika* "will differ from earlier atomic icebreakers by the fact that it is capable of working both in estuarial conditions which demand a small draft, and in deep waters which demand a large draft." Most people anticipate cost overruns, and funding already lags.

26. Beliakov, "Arkticheskie Vezdekhody."

27. In *Plutopia* (New York: Oxford University Press, 2013) Kate Brown chronicles the terrible human and environmental legacy of plutonium run amok in the United States and USSR.

28. "Rosatom Gears Up to Serve a Global Market," Aug. 21, 2012, http://www.rosatom .ru/en/presscentre/nuclear_industry/1f3573804c6d425bae9eafda0118feee.

29. Among the dozens of excellent compilations and reports that indicate the extent of nuclear waste and other problems of the Soviet Cold War legacy, see the Bellona Foundation reports: Nils Bøhmer, Aleksandr Nikitin, Igor Kudrik, Thomas Nilsen, Andrey Zolotkov, and Michael H. McGovern, *The Arctic Nuclear Challenge* (Bellona Foundation, 2001), and Igor Kudrik, Aleksandr Nikitin, Charles Digges, Nils Bøhmer, Vladislav Larin, and Vladimir Kuznetsov, *The Russian Nuclear Industry—The Need for Reform* (Bellona Foundation, 2004).

30. "The Russian Nuclear Industry," http://www.rosatom.ru/en/about/nuclear_industry /russian_nuclear_industry/, accessed Sep. 22, 2014.

31. On BAM, for example, see Christopher Ward, *Brezhnev's Folly: The Building of BAM and Late Soviet Socialism*. (Pittsburgh, PA: University of Pittsburgh Press, 2009). For an early variant of technological display, see Matthew Payne, *Stalin's Railroad: Turksib and the Building of Socialism* (Pittsburgh, PA: University of Pittsburgh Press, 2011).

32. On Atommash, see Paul Josephson, *Red Atom* (Pittsburgh, PA: University of Pittsburgh Press, 2005), pp. 97–108.

33. "Atomnaia Energetika Perezhivaet Nastoiashchii Renessans, i v Blizhaishie 15–20 Let Al'ternativy Ei ne Budet," April 15, 2010, http://www.atomic-energy.ru/news/2010 /04/15/10481.

34. Ibid.

35. "Informatsionnyi Sait Bilibinskoi AES," http://www.bilnpp.rosenergoatom.ru/, accessed Sep. 16, 2014, http://miss2011.nuclear.ru/en/contestants/?id=1, and Josephson, *Red Atom*, pp. 136–38.

36. I. V. Kurchatov, "Nekotorye Voprosy Razvitiia Atomnoi Energetiki v SSSR," *Atomnaia Energiia* no. 3 (195): 5–10, and "O Vozmozhnosti sozdaniia termoiadernykh reaktsii v Gazovom Razriade," in ibid., pp. 65–75.

37. Kurchatov, "Rech' Tov. I. V. Kurchatova na XX s"ezde KPSS," http://vkpb2kpss .ru/book_view.jsp?idn=002416&page=601&format=html.

38. The breeders would produce plutonium from a fertile blanket of ^{235}U and ^{238}U, the latter of which would transmute into fissile plutonium for future generations of reactors, and the problem of nuclear fuel would be solved.

39. Josephson, *Red Atom*, pp. 47–80.

40. "Iurii Kazanskii: Reaktor BN-800—Eto Vopros Liderstva Rossii," http://www .atominfo.ru/news/air288.htm, accessed July 7, 2014. Breeder reactors are opposed by nonproliferationists of all stripes and by Russian ecologists. See Ana Uzelac, "IAEA Backs Controversial Neutron Reactor Plan," *Moscow Times*, Nov. 11, 2000, http://www .themoscowtimes.com/sitemap/free/2000/11/article/iaea-backs-controversial-neutron -reactor-plan/257496.html.

41. Claire Bigg, "Amid Nuclear Scare, Russia Pushes Ahead with Controversial Floating Reactors," April 22, 2011, http://www.rferl.org/content/russia_pushes_ahead _with_controversial_floating_nuclear_reactors/9502474.html. See also Yevgenia Borisova, "Floating Nuke Plant Drawing Opposition," *St. Petersburg Times*, March 16, 2001, and http://prop1.org/2000/safety/970930ru.htm.

42. See A. Nikitin and L. Andreev, *Plavuchie Atomnye Stantsii* (Oslo: Bellona Foundation, 2011), http://bellona.ru/filearchive/fil_Floating-npps-ru.pdf, and Atominfo, "Zhiteli Peveka Odobrili Plan Razmeshcheniia Plavuchei AES," Nov. 16, 2013, http://www.atom info.ru/newsg/n0149.htm.

43. Bigg, "Amid Nuclear Scare."

44. Mikhail Antropov, "Atomnye Ambitsii Rossii," March 18, 2010, http://www.ntv .ru/novosti/188376.

45. Palmer, *Dictatorship of the Air*. On Russian hero building, see L. L. Kerber, *Tupolev* (St. Petersburg: Politekhnika, 1999); Nikolai Bodrikhin, *Tupolev* (Moscow: Molodaia Gvardiia, 2011); G. V. Novozhilov, ed., *Iz Istorii Sovetskoi Aviatsii: Samolety OKB imeni S.V. Il'iushina* (Moscow: Mashinostroenie, 1990); and P. Ia. Kozlov, *Velikoe Edinstvo: Dokumental'naia Poves'* (Moscow: DOSAAF SSSR, 1982).

46. Andrew L. Jenks, *The Cosmonaut Who Couldn't Stop Smiling: The Life and Legend of Yuri Gagarin* (DeKalb: Northern Illinois University Press, 2012).

47. For discussion of the political, social, and cultural importance of the space race to Soviet Russia, see Asif Siddiqi, *The Red Rockets' Glare: Spaceflight and the Soviet Imagination, 1857–1957* (Cambridge: Cambridge University Press, 2010).

48. Shaun Walker, "Putin Aims for the Stars with a £ 33bn Space Programme," April 12, 2013, http://www.independent.co.uk/news/world/europe/putin-aims-for-the -stars-with-33bn-space-programme-8570462.html.

49. "Is Vlad Keen on a Trip," April 12, 2014, http://www.dailymail.co.uk/news/article
-2602291/We-coming-Moon-FOREVER-Russia-sets-plans-conquer-colonise-space
-including-permanent-manned-moon-base.html.

50. "Russia Gearing Up for Launch of First Post-Soviet Rocket," *Moscow Times*,
June 27, 2014, http://www.themoscowtimes.com/business/article/russia-gearing-up-for
-launch-of-first-post-soviet-rocket/502608.html.

51. Simon Shuster, "Living and Dying with Russia's Soviet Legacy," July 12, 2012,
http://www.time.com/time/world/article/0,8599,2082637,00.html.

52. President of Russia, "Security Council Meeting on Long-Term State Policy in the
Aviation Sector," April 1, 2011, http://eng.kremlin.ru/news/1994.

53. Simon Shuster, "The Fatal Flight of Superjet-100," May 15, 2012, http://www
.time.com/time/world/article/0,8599,2114872,00.html.

54. "Russia Fears 110 Dead as Boat Sinks on Volga River," July 11, 2011, http://www
.bbc.co.uk/news/world-europe-14099637.

55. Will Stewart, "Undercover US Agents Brought Down Our New Superjet: Russia's
Extraordinary Claim About Crash Which Killed 45," May 24, 2012, http://www
.dailymail.co.uk/news/article-2149377/Undercover-US-agents-brought-new-Superjet
-Russia-s-extraordinary-claim-crash-killed-45.html#ixzz3E49vBJjq.

56. "Problem-Plagued Sukhoi Superjet Fails Take-Off at Moscow Airport," Feb. 25,
2013, http://www.reuters.com/article/2013/02/25/uk-problem-plagued-sukhoi-superjet
-fails-idUSLNE91O011201302225.

57. "United Aircraft Corporation," http://www.uacrussia.ru/en/, accessed Sep. 22,
2014.

58. "Soveshchanie o Formirovanii Goszakaza na Samolety Otechestvennogo Proiz-
vodstva," October 4, 2012, http://kremlin.ru/news/16596.

59. "Antonov, Ilyushin and Tupolev Fading Away," https://www.strategypage.com
/htmw/htairmo/20100614.asp.

60. Graham, *Lonely Thoughts*.

61. *Citizens United v. Federal Election Commission*, 558 US 310 (2010).

62. In *Wheel of Fortune* (Cambridge, MA: Harvard University Press, 2012), Thane
Gustafson tracks the power, politics, and uncertainties surrounding Russia's oil and gas
industry from the last years of Soviet power to the present.

63. "Gazprom Says Shtokman 2008–09 Budget Over $800 mln," Oct. 20, 2008,
http://in.reuters.com/article/2008/10/20/shtokman-budget-idINLK43254820081020.

64. "Russia's Giant Shtokman Gas Field Project Put on Indefinite Hold Over Cost
Overruns and Failed Agreements," Aug. 29, 2012, http://www.bellona.org/articles/articles
_2012/Shtokman_freeze.

65. Ol'ga Mordiushenko, Aug. 30, 2012, http://kommersant.ru/doc/2011009?isSearch
=True.

66. Alexander Panin, "Western Sanctions Could Damage One-Fifth of Russia's Oil
Production," *Moscow Times*, Sep. 21, 2014, http://www.themoscowtimes.com/business
/article/one-fifth-of-russia-s-oil-production-is-at-risk-due-to-sanctions/507474.html.

67. Alex McGrath, "Gazprom's Tower: Civil Society in the Venice of the North,"
http://petersburg.blogs.wm.edu/2011/10/04/gazprom%E2%80%99s-tower-civil-society
-in-the-venice-of-the-north-by/#_ftn1.

68. Ibid. McGrath wonders if the Gazoskreb had any connection with Vladimir
Tatlin's 1919 "Monument to the Third International," also "a spiraling mass of iron and
steel that exemplified Russian Constructivist architecture and was to stand in the city's
center," and also to rival and surpass the height of other European structures.

69. In an open letter to Governor Matvienko in June 2006, the Saint Petersburg Union of Architects wrote,

> The Construction of a high-rise building, which is sure to be visually connected to the historic center, we firmly believe, is absolutely unacceptable. . . . St. Petersburg is very uniform in height, so that its external appearance is dominated by the lines corresponding to the regularity of its plan. The low skyline of St. Petersburg makes its verticals particularly majestic, as they are almost always perceived against the sky. The preservation of the unique silhouette of steeples and domes is of great urban and spiritual importance. . . . The construction of a 300-meter tower will inevitably destroy the harmony of St. Petersburg's verticals, which has evolved for centuries, and will cause irreparable damage to the delicate silhouette of the city, practically making toys out of all the other verticals of the city . . . implementation of this construction will mean a complete break with St. Petersburg's urban-architectural tradition. (Alex McGrath, ibid., cited in Vishnevsky "Gazoskreb," 23)

70. UNESCO (the United Nations Educational, Scientific and Cultural Organization) officials rightly worry that the tower "might mar the historical panorama of the city's center and disrupt the architectural integrity of St. Petersburg." McGrath, ibid.

71. Ibid.

72. Ibid. Yury Volchok, architectural historian and professor at Moscow Architectural Institute, commented: "The situation with the Okhta Center repeats itself. The construction site was moved to the Primorsky District, which is not considered a part of the historical center, so it is easier to get the construction permit there. However, Russian legislation protects not only buildings but also city views." See "IuNESKO Mozhet Iskliuchit' Sankt-Peterburg iz Ob"ektov Vsemirnogo Naslediia," http://izvestia.ru/news /533615#ixzz24ScBI6f9.

73. "Lakhta Center," http://proektvlahte.ru/en/. "Okhta Center," http://www.ohta -center.ru/en/, has closed down.

74. Anti-Corruption Foundation, *Sochi 2014: Comprehensive Report*, http://sochi.fbk .info/en/.

75. Thomas Grove, "Russia's $50 Billion Gamble on 2014 Olympics," http://www .nbcnews.com/id/50892025/ns/business-world_business/#.USfABh3BLmc.

76. Boris Nemtsov and Leonid Martyniuk, *Zimniaia Olipiada v Subtropikakh* (Moscow 2013), http://www.nemtsov.ru/?id=718789.

77. On the monstrous cost to taxpayers of privately owned stadiums in New York City, see "As Stadiums Rise, So Do Costs to Taxpayers," *New York Times*, Nov. 4, 2008.

78. Boris Vishnevskii, "Vsem Mirom—dlia 'Gazproma'?" *Novaia Gazeta*, Oct. 22, 2012, http://www.yabloko.ru/publikatsii/2012/10/22. The government was able to go $400 million over budget to open a new Mariinsky Ballet and Opera Hall in May 2013, but critics generally find it to mesh harmoniously with the Petersburg skyline and architecture. See Louise Levene, "Mariinsky 2 Opens with a Seamless Gala," May 3, 2013, http://www.telegraph.co .uk/culture/theatre/dance/10034976/Mariinsky-2-opens-with-a-seamless-gala.html.

79. President Franklin D. Roosevelt, "The New Deal," July 2, 1932, http://www .danaroc.com/guests_fdr_021609.html.

80. President Franklin D. Roosevelt, Dedication of the Grand Coulee Dam, Washington, Aug. 4, 1934, http://digitalcollections.lib.washington.edu/cdm/singleitem /collection/panoram/id/4.

81. Roosevelt, "Address at Bonneville Dam," Sep. 28, 1937, http://newdeal.feri.org /speeches/1937c.htm.

82. President Bill Clinton, "Remarks [on the Human Genome Initiative]," June 26, 2000, http://www.genome.gov/10001356.

83. Ibid.

84. "President Bush Offers New Vision for NASA," Jan. 14, 2004, http://www.nasa.gov/missions/solarsystem/bush_vision.html.

85. Andrei Anishchuk, "Russia's Putin Restores Stalin-era Labor Award," May 1, 2013, http://www.reuters.com/article/2013/05/01/us-russia-putin-medal-idUSBRE9400HL20130501.

86. Fred Pearce, "Russia Reviving Massive River Diversion Plan," Feb. 9, 2014, http://www.newscientist.com/article/dn4637-russia-reviving-massive-river-diversion-plan.html, and Energy Resources Group, "Moscow Actively Considering Diverting Siberian River Water for Profit," Sep. 26, 2010, http://groups.yahoo.com/group/energyresources/message/125152.

CONCLUSION: What Have We Learned from This?

1. Regulation 536/2013, *Official Journal of the European Union*, June 11, 2013, L160/4–L160/5.

2. "EFSA Paves Way for EU High Fructose Corn Invasion," Oct. 23, 2013, http://anh-europe.org/news/efsa-paves-way-for-eu-high-fructose-corn-syrup-invasion.

3. Scott Sherman, "University Presses Under Fire," *Nation*, May 26, 2014.

4. Peter Dougherty as quoted in Sherman, "University Presses Under Fire," *Nation*, May 26, 2014.

5. Charles Richmond Henderson, "Are Modern Industry and City Life Unfavorable to the Family?" *American Journal of Sociology* 14, no. 5 (Mar. 1909): 672.

6. "Where Do You Take Your Phone," July 11, 2013, http://www.jumio.com/2013/07/where-do-you-take-your-phone/.

7. Brian S. McKenzie, "Public Transportation Usage Among US Workers: 2008 and 2009," *American Community Survey Briefs*, Oct. 2010, http://www.census.gov/prod/2010pubs/acsbr09-5.pdf.

8. Tanya Snider, "Census: American Bike Commuting Up Nine Percent in 2012," http://usa.streetsblog.org/2013/09/19/census-american-bike-commuting-up-nine-percent-in-2012/.

9. US Energy Information Administration, "How Much Gasoline Does the US Consume," May 2013, http://www.eia.gov/tools/faqs/faq.cfm?id=23&t=10.

10. Jason Szep, "Q and A with Amtrak President Alex Kummant," June 12, 2008, http://www.reuters.com/article/2008/06/12/us-usa-gasoline-trains-qa-idUSSIB27628520080612?sp=true.

11. David Leonhardt, "The Paradox of Corporate Taxes," *New York Times*, Feb. 2, 2011.

12. Tad DeHaven, "Corporate Welfare in the Federal Budget," *Policy Analysis* 703 (July 24) (Washington, DC: CATO Institute, 2012), p. 11, http://www.cato.org/sites/cato.org/files/pubs/pdf/PA703.pdf.

13. Ministerie van Verkeer en Waterstaat, *Cycling in the Netherlands*, http://www.fietsberaad.nl/library/repository/bestanden/CyclingintheNetherlands2009.pdf.

14. Martha Moore Trescott, "The Bicycle, A Technical Precursor of the Automobile," http://www.thebhc.org/publications/BEHprint/v005/p0051-p0075.pdf.

15. On this topic, see David Hounshell, *From the American System to Mass Production, 1800–1932* (Baltimore: Johns Hopkins University Press, 1984).

16. Peter Norton, *Fighting Traffic: The Dawn of the Motor Age in the American City* (Cambridge, MA: MIT Press, 2008).

17. Lewis Mumford, *From the Ground Up* (New York: Harcourt, Brace, and World, 1956).

18. Peter Samuel, "HOV Lanes Clogged with Hybrids—Complicate Toll Plan," Jan. 12, 2005, http://tollroadsnews.com/news/hov-lanes-clogged-with-hybrids-complicate-toll-plan.

19. Mayor Penalosa as cited in Jacob Koch, "Building More Roads Does Not Ease Congestion," Nov. 8, 2011, http://thecityfix.com/blog/building-more-roads-does-not-ease-congestion/.

20. See Thomas Hughes, *Rescuing Prometheus* (New York: Pantheon, 1998), ch. 5, on the engineering of "Big Dig."

21. David Kravets, "Motorist Claims Corporation Papers Are Carpool Passengers," Jan. 9, 2013, http://www.wired.com/2013/01/corporation-carpool-flap/.

22. See R. Arnott and K. Small, "The Economics of Traffic Congestion," *American Scientist* 82 (1994): 446, as cited in "Campaign for Sensible Transportation," "Why Widening Highway 1 Won't Work," May 12, 2002, http://sensibletransportation.org/pdf/fallacies.pdf.

Suggested Further Reading

INTRODUCTION: Technostories

Robert Friedel, *Zipper: An Exploration in Novelty* (New York: W. W. Norton, 1994).

Thomas Hughes, "The Evolution of Large Technological Systems," in *The Social Construction of Technological Systems*, ed. Wiebe Bijker, Thomas Hughes, and Trevor Pinch (Cambridge: MIT Press, 2012), pp. 45–76.

David Nye, *Consuming Power* (Cambridge, MA: MIT Press, 1998).

CHAPTER 1: The Ocean's Hot Dog

"The K Ration," http://www.usarmymodels.com/ARTICLES/Rations/krations.html.

Mark Kurlansky, *Cod: A Biography of the Fish That Changed the World* (New York: Walker, 1997). See also his *Frozen in Time: Clarence Birdseye's Outrageous Idea About Frozen Food* (2012), for middle school readers.

Eric Schlosser, *Fast Food Nation* (New York: Houghton Mifflin, 2001).

William Warner, *Distant Water: The Fate of the North Atlantic Fisherman* (Boston: Little, Brown, 1983).

CHAPTER 2: The Sports Bra

Nina Lerman, Ruth Oldenziel, and Arwen Mohun, eds., *Gender and Technology: A Reader* (Baltimore: Johns Hopkins University Press, 2003).

Londa Schiebinger, *The Mind Has No Sex? Women and the Origins of Modern Science* (Cambridge, MA: Harvard University Press, 1989).

Ruth Schwartz Cowan, *More Work for Mother: The Ironies of Household Technology from the Open Hearth to the Microwave* (New York: Basic Books, 1983).

Andrea Tone, *Device and Desires: A History of Contraception in the United States* (New York: Hill and Wang, 2001).

CHAPTER 3: Sugar, Bananas, and Aluminum Cans

Michael Adas, *Machines as the Measure of Men: Science, Technology, and Ideologies of Western Dominance* (Ithaca, NY: Cornell University Press, 1989).

Roy Macleod, ed., *Nature and Empire: Science and the Colonial Enterprise*, Osiris 15 (Chicago: University of Chicago Press, 2000).

Paolo Palladino and Michael Worboys, "Science and Imperialism," *Isis* 84, no. 1 (1993): 91–102.

James Scott, *Seeing Like a State* (New Haven, CT: Yale University Press, 1999).

CHAPTER 4: Mass-Produced Nutrition

Jacques Ellul, *The Technological Society* (New York: Knopf, 1964).
Marion Nestle, *Food Politics: How the Food Industry Influences Nutrition and Health* (Berkeley: University of California Press, 2003).
Michael Pollan, *Omnivore's Dilemma* (New York: Penguin, 2006).
Eric Schlosser, *Fast Food Nation* (New York: Houghton Mifflin, 2001).
Upton Sinclair, *The Jungle* (1906).

CHAPTER 5: Technology and (Natural) Disasters

Ulrich Beck, *Risk Society: Towards a New Modernity* (London: Sage, 1992).
Pete Daniel, *Deep'n As It Comes: The 1927 Mississippi River Flood* (New York: Oxford University Press, 1977).
John McPhee, *The Control of Nature* (New York: Farrar, Straus, Giroux, 1989).
Grigori Medvedev, *The Truth About Chernobyl* (New York: Basic Books, 1991).
Charles Perrow, "Disasters Evermore? Reducing Our Vulnerabilities to Natural, Industrial, and Terrorist Disasters," *Social Research* 75, no. 3 (Fall 2008): 1–20.

CHAPTER 6: Big Artifacts

Gabrielle Hecht, *The Radiance of France* (Cambridge, MA: MIT Press, 1998).
Thomas Hughes, *American Genesis: A Century of Invention and Technological Enthusiasm, 1870–1970* (New York: Viking, 1989).
Walter McDougall, *. . . the Heavens and the Earth: A Political History of the Space Age* (New York: Basic Books, 1985).
David Nye, *American Technological Sublime* (Cambridge, MA: MIT Press, 1994).
Albert Speer, *Inside the Third Reich* (New York and Toronto: Macmillan, 1970).

Index